MODELING MINERAL AND ENERGY MARKETS

MODELING MINERAL AND ENERGY MARKETS

by

Walter C. Labys
Professor of Resource Economics
Benedum Distinguished Scholar
West Virginia University

KLUWER ACADEMIC PUBLISHERS
BOSTON / DORDRECHT / LONDON

Distributors for North, Central and South America:
Kluwer Academic Publishers
101 Philip Drive
Assinippi Park
Norwell, Massachusetts 02061 USA
Telephone (781) 871-6600
Fax (781) 871-6528
E-Mail <kluwer@wkap.com>

Distributors for all other countries:
Kluwer Academic Publishers Group
Distribution Centre
Post Office Box 322
3300 AH Dordrecht, THE NETHERLANDS
Telephone 31 78 6392 392
Fax 31 78 6546 474
E-Mail <orderdept@wkap.nl>

 Electronic Services <http://www.wkap.nl>

Library of Congress Cataloging-in-Publication Data

A C.I.P. Catalogue record for this book is available
from the Library of Congress.

To Jane and the Families of TWA 800

CONTENTS

LIST OF FIGURES

LIST OF TABLES

PREFACE

This book provides a framework for analyzing and forecasting a variety of mineral and energy markets and related industries. Such modeling activity has been at the forefront of the economic and engineering professions for some time, having received a major stimulus following the first oil price shock in 1973. Since that time, other shocks have affected these markets and industries, causing disequilibrium economic adjustments which are difficult to analyze and to predict. Moreover, geopolitics remains an important factor which can destabilize crude oil markets and associated refining industries. Mineral and energy modeling, consequently, has become a major interest of energy-related corporations, mining and drilling companies, metal manufacturers, public utilities, investment banks, national government agencies and international organizations.

This book hopes to advance mineral and energy modeling as follows: (1) The modeling process is presented sequentially by leading the model builder from model specification, estimation, simulation, and validation to practical model applications, including explaining history, analyzing policy, and market and price forecasting; (2) New developments in modeling approaches are presented which encompass econometric market and industry models, spatial equilibrium and programming models, optimal resource depletion models, input-output models, economic sector models, and macro-oriented energy interaction models (including computable general equilibrium); (3) The verification and application of the models is considered not only individually but also in relation to the performance of alternative modeling approaches; and (4) The modeling framework includes a perspective on new directions, so that the present model building advice will extend into the future.

I would like to thank the many persons who have contributed to the completion of this work. The original stimulus to prepare an energy model framework was provided by the late David Wood who invited me to be a Visiting Scholar in the Energy Lab at the Massachusetts Institute of Technology during 1981-82. The resulting studies on energy modeling approaches were published in the Energy Lab working paper series. I also worked with Joel Clark in the Department of Material Sciences at the same Institute to produce an evaluation of mineral demand modeling, the latter of which was supported and published by the U.S. National Academy of Sciences. The results of both of these efforts were later published in *Mineral Economics*, by the American Institute of Mining Engineers. Bill Miernyk and Adam Rose have helped in my use of input-output models. And much of the research on optimization and programming models was performed in cooperation with Takashi Takayama and Chen-Wei Yang.

The present effort to update these modeling efforts was sponsored in part by the United States-Polish Maria Sklodowska-Curie Joint Fund in cooperation with Wojciech Suwala from the Mineral and Energy Economic Research Institute, the Polish Academy of Sciences, Krakow. Graduate students in the Natural Resource Economics Program at West Virginia University and in the Environmental and Resource Economics Seminar at the Institute for Advanced Studies in Vienna also participated by reading parts of the manuscript and offering suggestions for improvements. Finally, John Tilton of the Colorado School of Mines provided me with the opportunity in the spring of 1997 to prepare the final chapter on the future of modeling developments. Word processing assistance was given by April MacDonald, Stacia Rosenau and Eva M. Thomas, and Gloria Nestor helped with the graphics. Needless to say, the views expressed in this work do not necessarily reflect those of these individuals nor of any institutions with which I have been associated during the preparation of this work.

1 INTRODUCTION

Economists have found the modeling of mineral and energy markets and industries of considerable help in examining resource and energy related policies, national economic and environmental policies and in forecasting prices and quantities on the commodity markets of interest. Examples of such markets which have the potential to be modeled are given in Table 1. The concept of a model usually evokes an image of a complex, computerized system of mathematical or econometric equations providing detailed information concerning the operation of the process being modeled. In fact, models may be simple or complex, formal or mental, depending on the purposes for which a model is intended. Simple judgmental models may be most appropriate when monitoring the overall performance of a process. When more detailed information is required and/or when the model is used for complex decision steps, such as the choice of an optimal generation mix for an electric utility, then more complicated models are appropriate. The choice of theoretical structure, implementation methods, and the level of detail represent the art as distinct from the science of modeling. The initial step in considering any model, then, is to determine the appropriateness of the detail, theory, and implementation methods in relation to the purposes for which the model is intended.

The scope of mineral and energy system modeling ranges from engineering models of energy conversion processes (e.g., coal burning power plants) or components of such processes to comprehensive system models of a nation's economy in which the energy or mineral system is identified as a sector. Some of the models considered here are characterized by the coverage of various mining activities, fuel supplies, fuel demands, and by the methodology employed. Thus, the scope of the models reviewed includes addressing the supply and/or demand for specific processed commodities, analysis of intercommodity substitution and competition in a more complete systems framework, and analysis of the interrelationships between minerals, energy, the economy, and the environment.

Mineral and energy models are employed for both normative or descriptive analyses and predictive purposes. In normative analysis, the primary objective is to measure the impact on the system of changing some element of process that is exogenous, or independent, even in the model.

Table 1

PRINCIPAL MINERAL AND ENERGY COMMODITIES

Nonfuel Mineral Commodities		Fuel Mineral or
Metals	Nonmetals	Energy Commodities
Iron	Building	Fossil
Iron ore	Sand/gravel	Bituminous Coal
	Limestone	Lignite
Iron Alloys	Cement	Anthracite Coal
Manganese		Petroleum
Chromite	Chemical	Natural gas
Nickel	Sulfur	
Molybdenum	Salt	Nonfossil
Cobalt		Uranium
Vanadium	Fertilizer	
	Phosphate	Processed
Base	Potash	Electricity
Copper	Nitrates	Refined oil products
Lead		Synthetic fuels
Zinc	Ceramic	
Tin	Clay	
	Feldspar	
Light		
Aluminum	Defractory and Flux	
Magnesium	Clay	
Titanium	Magnesia	
Precious	Abrasive	
Gold	Sandstone	
Silver	Industrial diamonds	
Platinum		
	Insulant	
Rare	Asbestos	
Radium	Mica	
Beryllium		
	Pigment and Filler	
	Clay	
	Diatomite	
	Borite	
	Precious and	
	Gem diamonds	
	Amethyst	

Predictive models are used to forecast mineral or energy supply and/or demand and attendant effects over a particular time horizon. Most models have both normative and predictive capability, and a partition of models into these classes can be misleading. Whenever such a classification is used here, it is intended only to identify the primary objective of the model.

Geographical detail appropriate for a given model again depends on the purposes for which the model has been designed. For example, a model of energy flows in a particular production process is specifically related to the plants in which that process operates. Such a model has no geographical dimension. However, a model of utility electricity distribution has a very explicit regional dimension, defined by the utility service area being modeled.

Treatment of uncertainty or risk in a model is an important distinguishing characteristic. Uncertainty may arise because certain elements of the process to be modeled are characterized by randomness, because the process is measured with uncertainty, and because certain variables used as inputs to the model may themselves, be measured with uncertainty. The methods for dealing with these problems are important in evaluating predictive capability and in validating a model.

The validation of normative models is quite different from that of predictive models. Since normative models deal with how a mineral or energy system should develop given an objective, the issues of validation deal more with the realism of the description of the energy system and the accuracy of its input parameters. For predictive models, validation includes evaluation of both the model's logical structure and its predictive power. Three levels of predictive capability may be identified including ability to predict: (1) the direction of a response to some perturbing factor (e.g., a decrease in GDP due to a fuel supply curtailment), (2) the relative magnitude of a response to alternative policy actions or perturbing factor; and (3) the perturbing factor itself. Validation against the requirement of the first two levels is a minimum requirement, and a model may be quite useful even if it cannot be validated at the third level. At both the second and third level, validation of a conditional form is usual, and restrictions on the perturbing factors and their range of availability must be specified. Intervention analysis can be used in this regard. Perturbing events outside the scope of the model, such as military interventions, of course, must also be taken into consideration in evaluating predictive capability.

Mineral and energy models are formulated and implemented using the theoretical and analytical methods of several disciplines including engineering, economics, econometrics, time series analysis, operations research, and management science. Models based primarily on economic theory tend to emphasize the behavioral characteristics of decisions to produce and/or to utilize minerals or energy, whereas models derived from engineering concepts tend to emphasize the technical aspects of these processes. Process models are usually implemented using programming techniques and/or methods of network and activity analysis, whereas the behavioral models use econometric methods. Behavioral models are usually oriented toward forecasting uses, whereas process models tend to be normative. Time series models also are applied to forecasting, based on the history of a time series itself.

Recent modeling efforts (as discussed later) evidence a trend toward combining the behavioral and process approaches to mineral and energy modeling in order to provide a more comprehensive framework in which to forecast the conditions of future markets under alternative assumptions concerning the emergence of new production, conversion, and utilization technologies. In part, this trend is the result of recognizing that formulating and evaluating alternative natural resource policies and strategies requires an explicit recognition of technical constraints.

Mathematical programming has been used in mineral and energy modeling to capture the technical or engineering details of specific energy supply and utilization processes in a framework that is rich in economic interpretation. In mathematical programming, series of activity variables are defined representing the levels of activity in the specific processes. These can be arranged in a series of simultaneous equations representing, for example, demand requirements, supply capacities, and any other special relationships that must be defined to typify technical reality or other physical constraints that must be satisfied. An objective function to be minimized or maximized must be specified (usually cost, revenue, or profit); there are many algorithms available to solve very large problems. The linear programming (LP) technique has been used far more than other mathematical programming methods because of its efficiency in solving large piecewise linear or step function approximations. Linear complementarity, nonlinear (e.g., quadratic), variational inequality and dynamic programming techniques are also used for special purposes, but more often these methods exemplify market conditions which characterize economic market behavior.

Mathematical programming also can be associated with a dual problem formulated in terms of prices as compared to quantities. The solution to the quantity optimization problem yields both the optimal activity levels in physical terms and the prices (shadow) that reflect the proper (sometime marginal) valuation of physical inputs to the real process represented by the model. Important information concerning the economic interpretation of the solution is provided. While the LP technique provides a natural link between process and economic analysis, other techniques such as the calculus of variations and Lagrange multipliers are generally classified as normative techniques since they presume the existence of an overall objective such as cost minimization or profit maximization. It is possible to construct multi-objective criteria as some weighted combination of objectives, and indeed, some objectives such as environmental control can be expressed through special constraint equations in the model. Nevertheless, the validity of this technique as a predictive tool depends on its ability to capture and represent the objective of the players in various sectors of the mineral or energy system and in those sectors of the national or regional economy that affect these sectors. The technique is normative in that it determines optimal strategies to achieve a specific objective with a given set of constraints.

Econometrics is concerned with the empirical representation and validation of economic theories. The principle method of constructing related econometric models is to specify, estimate, and simulate a system of equations. This system can reflect directionality that is recursive or simultaneous in nature. Each of the equations is based on regression analysis. Such equations usually utilize economic theory to specify the relationships existing between the dependent and certain independent variables. Estimates of the related parameters then follow by combining the economic model with a statistical model of measurement and stochastic errors. Statistical distribution theory is also employed in performing hypothesis tests on the estimated parameters. Model validation is based both on such hypothesis testing, and on the predictive performance of the model, when forecasts and actual observations of the dependent variables are compared. Such predictive tests are most persuasive when applied outside the sample observations used in estimating model parameters.

Time series analysis can be considered as a particularized domain of econometrics. That is, the modeling equations are based on statistical theory but the generating processes underlying the time series variable to be modeled are more important than any economic explanatory factors. While this statement reflects the application of univariate time series models, today the attention devoted to multivariate time series models has permitted the modeling of economic inter-relations between time series variables. Econometric and time series methods are both used in modeling behavioral and technical processes. Behavioral processes are characterized by a decision-making agent hypothesized to adjust behavior in response to changes in variables outside his direct control. For example, one could hypothesize that a household might distribute its expenditures between energy and other types of goods and services on the basis of its income and wealth, the relative prices of energy, and the other products it consumes, and that this distribution would be consistent with some household objective function.

Technical process models are characterized by purely technical relations. An example would be the production of a firm in which maximum potential output is a function of the quantities of inputs available such as capital, labor, energy, and other material inputs. Given a suitable functional form for this relationship and observations on capacity output and associated inputs, econometric methods can be used to estimate the parameters of the relation. Alternatively, a technical relation might be used to derive behavioral relations concerning the firm's demand for input factors; for example, a firm could choose cost-minimizing combinations of inputs to produce a given output level. Finally, technical processes can be described by time series models such as in the case of the logistic growth function.

Econometric methods and engineering/process methods are sometimes alternative approaches for modeling technical processes. An example of two contrasting approaches to modeling the supply of electricity in the United States is provided by the work of Griffin (1974) and Baughman and Jaskow (1974). Griffin used an econometric approach while Baughman and Jaskow employed an engineering/process approach.

Interindustry or input-output methods are frequently employed in mineral and energy modeling, primarily for descriptive purposes. The interindustry flow table may be converted into a coefficient table measuring the quantity of an input required from one sector per unit of output for another sector. The coefficient matrix represents a model of the production process. This technique provides a means of linking technical coefficients relating input requirements (e.g., mineral or energy) per unit of output with behavioral models of demand for primary factors of production (capital and labor), and the demand for final goods and services. Thus, the interindustry framework provides a natural bridge between programming and econometric models of resource/economy interactions. The best early example of such an integrated model was that of Hoffman and Jorgenson (1977). The input-output approach has also been employed in energy studies by converting the inputs from the energy sector to other industry sectors from dollar flows into energy units, such as the British thermal unit (BTU).

Although any attempt to organize the above modeling approaches into distinct categories is somewhat arbitrary, the following taxonomy has been selected to highlight the scope of the major methodologies:

(1) Economic sector models including time series models which describe the supply, demand or prices for specific minerals, fuels or energy forms;

(2) Econometric market or industry models which include supply, demand and price aspects of market adjustments for individual or related mineral and energy commodities;

(3) Spatial equilibrium and programming models which apply programming algorithms to describe the distribution of demand and supply in a spatial or process context;

(4) Resource exhaustion forms of optimization models which describe how firm or industry cartels might establish prices to achieve optimal resource allocation over time;

(5) Input-output models which explain the flows of mineral or energy goods and services among industries in establishing outputs and demands;

(6) Economy interaction models which link the mineral and energy sectors with the overall economy, sometime integrating two or more modeling methods;

(7) Resource balance models which yield an empirical view of future energy demand and supply equilibria;

(8) And transition models which recently have been constructed to explain the transition of mineral markets from planned to competitive systems in the Eastern European nations.

The background for the presentation of these methods appears principally from studies by Labys and Wood (1985), by Labys, Field and Clark (1985), and by Labys and Yang (1991, 1997).

2 ESSENTIAL STEPS TO MODELING

Before presenting each of the above methodologies, it is important to understand the steps essential for mineral and energy modeling. Developing and applying a mineral or energy model normally requires the same "tool kit" that is used for constructing a typical commodity model, i.e. see Labys (1973, 1975) and Labys and Pollak (1984). Such a model consists of a number of components which reflect various aspects of demand, supply, trade and price determination. Each of these components, in turn, embody basic theories of economic behavior and/or transformation processes. What gives a mineral or energy model its particular configuration is the kind of market system it attempts to emulate. There is no such thing as an all-purpose commodity model. Each model must describe aspects of mineral or energy behavior which are peculiar to the system of interest.

Another aspect of mineral and energy models is that they involve a specific modeling procedure, as illustrated in Figure 1. The development of a model normally begins with an identification of the modeling problem to be solved. This is important since the modeling purpose must be clearly established from the outset. Modeling purpose normally leads to the selection of a modeling methodology that will effectively solve the problem at hand. In the next chapters, we will consider just how this selection takes place.

The most extensive modeling activities take place at the next stage. The model must be specified following the required modeling procedure. This requires a process which represents an interaction between selecting appropriate economic and engineering theories and establishing a data base for model implementation and application. This, in turn, provides the base for model estimation and calibration. In the context of mineral or energy models, parameters derived from econometric models depend on statistical methodologies, while parameters employed in engineering systems normally depend on engineering experiments or on judgmental observation.

In either case, the outcome of the next step, indicated as model validation and refinement, requires two properties: first that the model's parameters are demonstrable in accord with the underlying theory, and secondly that the model's output replicates reality closely. This procedure depends not only on the use of parametric and nonparametric validation

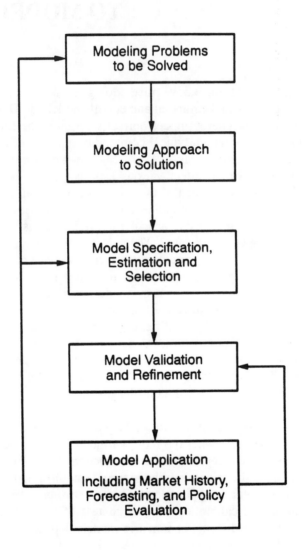

Figure 1

PROCEDURE FOR DEVELOPING AND APPLYING
MINERAL AND ENERGY MODELS

validation tests, but on other considerations as well. The overall process of validation is described in chapter 11.

The final step of modeling involves applying the model to solve the problem at hand. This process normally consists of explaining market history, of analyzing selected policies, or of forecasting future paths of the model's dependent variables. This step is the most important of all, and its feedback connection with previous steps should be kept in mind, as shown in Figure 1. If the application does not prove effective, then the modeler must return to an earlier stage to identify and to reconstruct the offending model components by respecification, re-estimation and calibration. In the following chapters, the particular character of each of these steps should become more clear. Let us now turn to the major modeling methodologies outlined previously.

3 ECONOMIC
SECTOR MODELS

The most elementary forms of econometric models are not models in the complete sense. Rather they are components of models or are single sector models that relate to one particular aspect of a mineral or energy market or industry as a whole. Typically, model components or equations in this category focus on the price, the supply, the demand or the trade aspect of a market. The underlying methodologies used for the components can be statistical equations which embody some complex time-series process or econometric regression equations which feature dependence on a set of economic and technological factors of an explanatory nature.

Statistical models based on time series analysis assume that the future value of a single mineral or energy variable such as demand or prices can be predicted by using the past representation of that variable. That is, by examining the underlying time series generating process of a variable, one can find a time-dependent model based on statistical theory which will permit an explanation and prediction of the time behavior of that variable. The actual identification of that process itself is a step by step procedure which eventually will lead to model specification, e.g. see Cromwell, Labys and Terraza (1994a).

Most such models employ linear and/or nonlinear temporal forms, e.g. see Granger and Newbold (1986) or Granger and Terasvirta (1993). As an example, the time dependent energy variable X_t can thus be expressed as some function of time trend T_t.

$$X_t = b + aT_t \tag{3.1}$$

This same trend relation can also be formulated nonlinearly such as in the case of a simple exponential in mean.

$$X_t = b + \exp(\alpha T_t) \tag{3.2}$$

Nonlinear specification can also be formulated in terms of variance specifications, such as in ARCH or GARCH models.

More typically X_t is explained based on a structure more complex than a simple curve. This has been accomplished historically by decomposing the X_t series into separate component models reflecting trend, cyclical, seasonal and irregular behavior. Today, this procedure has been simplified by using the autoregressive-moving average modeling approach. One begins by first selecting an autoregressive model which generates X_t as direct function of its past values. The simplest is that of first-order autoregression (AR)

$$X_t = \alpha \, X_{t-1} + e_t \qquad (3.3)$$

where e_t is zero-mean and randomly distributed. The more general representation is given by

$$X_t = \sum_{j=1}^{p} \alpha_j \, X_{t-j} + e_t \qquad (3.4)$$

where a number of lags j up to p can be selected.

Prediction may also be made by smoothing the behavior of X_t in the form of a moving average (MA)

$$X_t = \sum_{j=1}^{q} b_j \, e_{t-j} \quad where \quad b_o = 1 \qquad (3.5)$$

In this case X_t is expressed as a weighted average of past values of e_t with up to q values. This method simply follows the error process of X_t taken over q periods and uses this as the forecast for the next period.

When the data underlying X_t permit, one can then combine the above two approaches to form the mixed autoregressive-moving average model (ARMA), e.g. see Cromwell, Labys and Terraza (1994a).

$$X_t = \sum_{j=1}^{p} \alpha_j \, X_{t-j} + \sum_{j=1}^{q} b_j \, e_{t-j} \qquad (3.6)$$

When this model involves integration after X_t has been differenced to be stationary, it becomes known as the autoregressive integrated moving average model of order p, d and q

$$a(B)(1 - B)^d X_t = b(B) e_t \qquad (3.7)$$

where B is a stationary operator, i.e. ARIMA (p,d,q). Because computer software has been developed to deal easily with the related estimation and forecasting problems, this method has proven popular for explaining and predicting economic variables, e.g., see Box and Jenkins (1970) and Nelson (1973). It is particularly appropriate when dealing with the prediction of mineral or energy variables that are observed on a short term basis, i.e., quarterly or monthly. More recently, the integer d can be fractional and the model becomes ARFIMA. Kouassi et al. (1996) used this approach to forecast mineral prices, and more recently Beck and Solow (1994) have used a Bayesian version to forecast nuclear power supplies. In addition Uri (1979) has used ARMA models to predict electricity demand and Ringwood et al. (1993) have extended such demand forecasts by using ARMAX models.

Consideration of a wider set of explanatory factors for a given mineral or energy variable introduces two additional concepts to the above analysis. First, it is necessary to examine and make use of any theory that postulates relationships determining the dependent variable. Second, one can utilize this theory to include one (simple-regression) or several (multiple-regression) explanatory variables. The multivariate extension of the autoregressive model given above would analyze the relationship between such variables and is known as the vector autoregressive (VAR) model, e.g. see Cromwell et. al. (1994) and Sims (1980). The VAR representation of a k dimensional variable system can be given by

$$Y_t = C + A_1 Y_{t-1} +,...,+ A_p Y_{t-p} + e_t \qquad (3.8)$$

where Y_t is a k x 1 vector of variables, A are (k x k) coefficient matrices, C is a k x 1 vector of deterministic components, and e_t is a 1 x k vector of white noise residuals, i.e. Σ_e is a k x k positive-definite diagonal matrix. Other possible deterministic components include a constant term or dummy variables that represent discrete shifts in the relationships at a specific point in time.

In this respect the above model can be called a pure or unconstrained VAR model to distinguish it from a modified VAR model in which certain coefficients are restricted to be zero. While the choice of coefficients can be determined using the Bayesian vector autoregressive on some other restrictive approach, one can also employ an alternative approach in the form of a modified VAR in which particular variables, so called driving variables, are permitted to influence other variables; at the same time these variables are free of feedback. In a VAR model, variables are normally not specified as being either endogenous or exogenous. This permits one to move from a general to a more specific model representation, allowing the final model to capture the available time series information. Such an approach makes the VAR model particularly suited for mineral and energy forecasting applications.

Because of the difficulty of direct interpretation of the coefficient matrices A, Sims (1980) has suggested the use of the moving-average (MA) representation of the model

$$Y_t = \sum_i^{\infty} \Phi_i\, e_{t-1} + e_t \qquad (3.9)$$

where Φ_i are the k x k matrices with summation from 0 to ∞. The individual matrices Φ_i are a result of the following expression

$$\Phi_i = \sum_j^{s} \Phi_{i-j} A_j$$

where $\Phi_0 = I_k$, the k x k identity matrix. The summation on j is from 0 to s, where s is the number of periods of the response.

It is preferable to orthogonalize e and to interpret the coefficients as the resulting MA representation; thus, a lower triangular matrix, denoted by p, with diagonal elements is chosen such that $pp = \Sigma_e$. This process can be accomplished through the use of a Choleski decomposition. Define $v_t = p^{-1} e_t$ then $E\ [v_t v_t\ ] = I_k$. The model can now be rewritten as

$$Y_t = \sum_i \theta_i\, v_{t-i} \qquad (3.10)$$

where $\theta_i = \Phi_{-1}\, p$. Each parameter θ_{ijh} (h = 1,.....,m) traces the response over time of variable i following an innovation to variable j. Consider the specification that the matrices are upper triangular with the first vector representing the dependent variables and the second the independent or driver variables. In this case, one can utilize impulse response functions which permit the examination of the response of a mineral or energy variable to a shock in the autoregressive representation of any driver variables.

Since in most estimation contexts the variance-covariance matrix is not diagonal, the ordering of the variables is important with respect to deriving the Choleski decomposition. Thus, it has been assumed that most of the correlation between the hypothetical dependent and other variables can be attributed to the other independent variables or the driver variables.

Another analytical tool that can be employed is that of variance decomposition. The proportions of the h-step forecast error variance of variable i, accounted for by innovations in variable j, can be defined by $w_{ij,h}$

$$ W_{ij,h} = \Sigma_i (s_{k'} \, \theta_i \, s_j)^2 \, / \, \Sigma_i \, s_k \, \Phi_i \, \Sigma_e \, \phi_{i'} \, s_k \qquad (3.11) $$

where s_k is the kth column of I_k and Σ_i ranges from $i = 0$ to $h = 1$. One should remember that VAR models do have strong forecasting abilities, which makes them useful for forecasting mineral and energy variables.

The actual construction of a VAR model for mineral and energy purposes is a rather complex process. Similar to the specification of univariate time series models, a step by step procedure employing tests of process behavior identification is required. As described in Cromwell, et al. (1994b), tests of joint stationarity and independence, cointegration and causality are required. The latter tests determine whether two variables have a long run relationship over time or whether one variable Granger-causes a second variable over time. These tests are sufficiently important that they have been employed (independent of VAR considerations) to determine possible relations between mineral and energy variables. In particular cointegration tests have been applied between energy demand, energy prices and gross domestic product to better interpret energy demand elasticities, e.g. see studies by Bentzen (1994), Bentzen and Engsted (1993 a and b), Hunt and Manning (1989), Harvie and Vantloa (1993), and Yu and Jin (1992).

Regarding causality tests, Labys and Maizels (1993) have evaluated the extent to which mineral and energy price fluctuations have affected macroeconomic adjustments in the major OECD countries. More formal VAR models related to such identification tests have taken several directions. For example, Labys, Murcia and Terraza (1996) have constructed a VAR model to forecast movements in the Rotterdam oil spot market, while Stern (1993) has used a VAR model to analyze how energy use affects economic growth in the United States. Other VAR applications have used impulse tests to evaluate the impact of the petroleum price shocks on several economies, e.g. see Bohi (1989) and Burbridge and Harrison (1984).

DEMAND MODELS

Many econometric modeling efforts of a multivariate nature describing the energy sector have focused upon the demand for a single energy input in one particular use. Such models are used principally to provide an analysis of the basic determinants of demand as well as to forecast demand. The theory employed for this analysis derives from microeconomics. For example, demand D for a fuel can be expressed as a function of its price P, the price of competing fuels PC, and energy utilizing activity A or income GDP.

$$D_t = \beta_o + \beta_1 P_t + \beta_2 PC_t + \beta_3 A_t + e_t \qquad (3.12)$$

Here the variables D, P, PC and A are assumed to be independently distributed; the disturbance term e is zero-mean and normally distributed. Regression methods are used to estimate the coefficients, e.g. see Labys (1973) or Pindyck and Rubinfield (1997). When the variables are expressed in log form, the regression coefficients directly measure the direct price elasticity, the cross-price elasticity, and the income elasticity of demand, respectively.

Because economic sector equations are usually developed singly rather than in a more complex model form, they generally do not have broad policy applicability. Some examples of energy demand applications include that of Taylor (1975) who has surveyed and evaluated econometric equations of the short- and long-term demand for electricity in the residential and commercial sectors. Birol and Guerer (1993) have estimated demand equations for transport fuel requirements in developing countries. And Sweeney (1975) has developed econometric equations for the demand for gasoline in order to support analysis of energy conservation policies affecting automobiles.

SUPPLY MODELS

Mineral and energy supply can be explained using the above form of univariate or multivariate regression models or by engineering or accounting techniques which result in appropriate supply curves. These supply curves are also expressed with cost rather than price on the vertical axis and with supply on the horizontal axis cumulating to describe total reserves. The construction of supply curves with reserves depending on geological characteristics has been fostered by Harris (1984) in the context of mineral resource appraisal models. The fabrication of supply curves based on cost variables has been developed by Davidoff (1980), Foley and Clarke (1981), Torries (1988) and others.

According to Torries, a cost model is prepared for individual mines and that specification can be used to derive a cost curve or supply curve for a total industry. This reflects economic theory that an industry supply curve can be constructed from the summation of the marginal cost curves of individual firms as long as the actions of one firm do not affect the costs or actions of any other firm in the industry. For graphic simplicity, one assumes that the average and marginal cost curves of individual firms are flat up to the point of plant capacity and are nearly vertical thereafter. In a competitive industry, it is then possible to construct an industry stepwise supply curve by summing the marginal cost curves of the individual firms across the output range (horizontal axis).

Production costs of the firm can be classified in many ways but mainly as being either fixed or variable. Under this simplistic classification, production decisions of the firm will be based on average and marginal variable costs. The components of these costs can vary widely. In the case of a mine, smelter or refinery, the following cost categories are often employed: (1) labor, (2) energy, (3) materials and supplies, (4) local taxes and insurance, (5) transportation, (6) royalties, (7) direct operating costs, (8) byproduct credits, (9) net production costs (sales and administrative costs, interest, other cash costs), (10) cash breakeven costs (initial capital recoupment, return on investment), and (11) total costs.

It should be noted that these supply curves can be combined with demand curves. From the intersection of the demand curve and the cumulative marginal variable cost curve, an indication of commodity price can be determined, with the assumption that the industry is competitive. The price of a mineral is determined by aggregate supply, demand, and the highest cost producer required to fill demand. Note that producers with variable costs greater than some maximum are shut down while those firms operating with total costs greater than that level are not recovering their fixed costs. If the return on invested capital is considered to be a portion of fixed cost, these operating firms are making less than expected profits. On the other hand, those firms operating with total costs less than the maximum are making more than expected profits and are collecting economic rents.

Torries (1988) suggests that in reality two supply curves exist instead of one. Plants that close because of low prices do not immediately reopen as soon as prices increase to the point of equaling average variable costs. Instead, managers want assurance that prices will remain high for a sufficiently long period of time before the plant is reopened. In equation form, this can be represented by

$$P^* > AC + PR \qquad\qquad (3.13)$$

where:
P* = price necessary for reactivation
AC = average variable cost
PR = f' (E*, c*) - m
E* = Price premium determined by managerial expectations about
        future higher prices
c* = reactivation costs
m = maintenance and other ongoing costs when shut down

In the case that m<f' (E*, c*), the price necessary for reactivation will be higher than average variable cost.

Likewise, a manager will not close a plant until he is assured that prices will remain low for some sufficiently long period of time, considering closing and maintenance costs. In equation form this can be represented by

$$P < AC - PC \qquad\qquad (3.14)$$

where:
P   = price necessary for shutting down
AC = average variable cost
PC = f (E, c) + m
E   = price discount determined by managerial expectations about
        future lower prices
c   = closing costs
m = maintenance and other ongoing costs when shut down

When m < f (E, c), the price necessary to induce closing down is less than average variable costs. Other ongoing costs associated with closing a plant include contract cancellation penalties, loss of market share, loss of foreign exchange earnings and costs of unemployment.

The interest in constructing industry supply curves has existed for sometime, since such information is essential for evaluating break-even points, price structures and the profitability of mineral and energy firms. This interest was heightened during the 1970s when inflation and structural changes in world economies caused price forecasting as an input into evaluating new mineral projects to be a dubious exercise. Since 1980 a number of studies have been undertaken, attempting to derive costs of individual mining firms and the industry cost or supply curves. The best known of these is the supply analysis model, a minerals availability system methodology of the US Bureau of Mines, e.g. see Davidoff (1980). This system has been applied to a number of mineral industries including copper, manganese, nickel, phosphate, molybdenum, lead and zinc, in which production costs of individual mineral producers in the world were estimated. These analyses resulted in the estimation of long-run industry supply curves for each mineral commodity. Other similar studies have been completed for copper by Foley and Clark (1981), for molybdenum by Kovisars (1982), for nickel by Torries and Martens (1985), for uranium by Kovisars (1982b), and for coal by Price (1984).

Determining the costs of individual mineral producers and constructing an industry supply curve from the aggregate costs of individual producers is fraught with practical as well as theoretical difficulties. Mostly, this process produces good results when properly applied and can be used to identify and solve many practical problems involving availability and prices of mineral and energy commodities.

PRICE MODELS

The explanation and forecasting of mineral and energy prices also has been based on the above univariate and multivariate methods. The type of models to be constructed depends on whether the researcher is interested in long run as compared to medium run or short run price behavior. The modeling of long run behavior involves basic linear or nonlinear trend models as shown earlier. The explanation of medium run behavior can involve models capable of generating some form of price cycles; examples include ARIMA, ARFIMA, exponential smoothing, VAR or structural time series models. The deciphering of short run behavior often is concerned with stochastic or random processes, such as those associated with the discovery of futures price movements.

The background to constructing and forecasting mineral and energy price behavior is too extensive to review here. First of all, some of this work has been directed simply to the discovery of the time series generating processes underlying the price behavior. Among classical studies using special methods such as spectral analysis are Labys and Granger (1970) Labys, Elliott and Rees (1971), and Slade (1981). Today more emphasis has been placed on discovering the order or fractional order of integration using special testing procedures, e.g. see Barkoulas, Labys and Onochie (1997, 1998). Studies such as Giraud (1995), Greenberg and Conway (1989) and Plourde and Watkins (1995) have employed less robust methods to analyze crude oil price movements.

The second direction of this work has been to construct models more of an economic nature to explain mineral and energy price behavior. Some studies such as Chu and Morrison (1984) have been longer-term and more macroeconomic in their approach. Studies which concentrate more directly on the use of time series models are numerous. More recently Labys and Kouassi (1996) have applied the structural time series approach to discover stochastic cyclical behavior in these prices. Special interest in electricity pricing can be found in works such as Munasingha and Warford (1981), Schweppe et al (1988) and Turvey (1968). Other studies such as those presented in Winters and Sapsford (1990) employ models which are larger scale in nature and thus closer to the econometric models discussed in the next sections.

Finally, the third direction of price sector modeling has been to forecast both levels and turning points of mineral and energy price behavior, e.g. see Driehuis (1976). Such price forecasting exercises have long been conducted by the World Bank, as described by Duncan (1984). A review of just how different univariate and multivariate methods can be applied to minerals and metals prices was produced by the United Nations (1984) and also is useful for present purposes. Finally, Kouassi et al. (1996) among others have shown how more modern univariate methods can be applied to these same prices. For a criticism of the forecasting performance of many of these models see Gerlow, Irwin and Liu (1993).

4 ECONOMETRIC MARKET OR INDUSTRY MODELS

This chapter emphasizes the major econometric approaches to modeling mineral and energy markets. The first two sections feature forms of market models. The presentation then continues with what have become known as econometric demand and supply models. Energy conservation models are discussed in the closing section.

COMPETITIVE MARKET MODELS

The most basic type of model from which econometric and other mineral or energy modeling methodologies have developed is the competitive market model. Such a model initially neglects market imperfections and assumes that commodity demand and supply interact to produce a price level reflecting competitive market conditions. Such a model may consist of a number of combined regression equations, each explaining separately, a single market or sector variable, as described above. Market models or the equivalent industry models are applicable to all mineral or energy production and use categories. Their greatest utility is in providing a consistent framework for planning industrial expansion, forecasting market price movements, and studying the effects of regulatory policies. The basic structure of such a model typically explains market equilibrium as an adjustment process between demand, supply, inventory and price variables, e.g. see Labys (1973) or Labys and Lord (1992). Such a structure has been described in Figure 2. Most simply it would consist of the following equations:

$$D_t = d(D_{t-1}, P_t, P C_t, A_t, T_t) \qquad (4.1)$$

$$Q_t = q(Q_{t-1}, P_{t-\theta}, N_t, Z_t) \qquad (4.2)$$

$$P_t = p(P_{t-1}, \Delta I_t) \qquad (4.3)$$

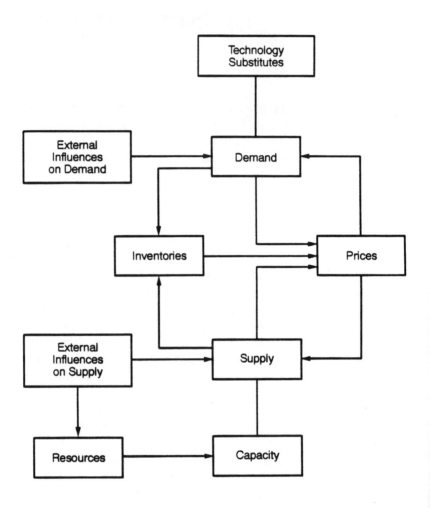

Figure 2

BASIC ECONOMETRIC COMMODITY MODEL

$$I_t = I_{t-1} + Q_t - D_t \qquad (4.4)$$

where:

D	=	Demand
Q	=	Supply
P	=	Prices
PC	=	Prices of substitutes
P $(-\theta)$	=	Prices with lag distribution
I	=	Inventories
A	=	Income or activity level
T	=	Technological factors
N	=	Resource factors
Z	=	Policy variables influencing supply

Commodity demand is explained as being dependent on prices, economic activity, prices of one or more substitutes and possible technological influences. Other possible influencing factors and the customary disturbance term are omitted here and below to simplify presentation. Accordingly supply would depend on prices as well as the underlying production factors, for example, such as geology or resource exhaustibility, and a possible policy variable. A lagged price variable is included since the supply process is normally described using some form of the general class of distributed lag functions. Commodity prices are explained by changes in inventories, although this equation is sometime inverted to explain inventory demand. The model is closed using an identity which equates inventories with lagged inventories plus supply minus demand. The utilization of this model involves further specification, estimation and simulation (Labys, 1973, 1978). Some experimentation has also taken place using system dynamics (Ruth and Hannon, 1997;Meadows, 1970)

Applications to energy markets have not been extensive because of the difficulties of dealing with regulatory policy and non-competitive influences on market behavior. Verleger (1982, 1993), however, has shown that it can be applied to explain disruptive shortages. This model links together econometric equations for oil spot prices, consumer demand, and shortage conditions. MacAvoy and Pindyck (1975) have built an econometric model of the natural gas industry which has been used extensively to analyze the effect on the industry of federal regulation of the wellhead price of gas and of permissible rates of return for the pipeline industry. Labys, et al. (1979) have modeled the U.S. coal market using this approach to forecast future levels of coal demand, supply, prices and inventories. Most recently, Trieu (1994) et al. have reported their modeling of the world spot uranium market.

The development of econometric models suitable for analyzing mineral markets possessing competitive behavior has been wide-spread. One of the first was the Desai (1966) tin model which explained tin price fluctuations on a world basis. The model was disaggregated on the demand side into three regions: the United States, the OECD country total, Canada, and the rest of the world. The total demand for tin in the former two regions was further disaggregated according to two end-use categories (tin-plate and non-tin-plate) to capture more accurately the influence of the relevant activity variables and technological changes in the end-uses. The immediately relevant variables relating to the use of tin for tin-plate and non-tin-plate were linked with larger macro-variables, such as GNP and industrial production. This model was used to study how various conditions within the tin market could be improved to stabilize the revenues of developing tin exporting countries. Policy analysis employing stochastic simulations aided in this respect. Other simulations carried out were aimed at investigating the possibility of reducing fluctuations in the prices and revenues received by tin producers, through the operation of an international buffer stock and the restriction of output by the International Tin Council. The basic model structure as well as the model purpose were later expanded by Smith and Shenk (1979) who compared buffer stock operations by the International Tin Council with strategic stockpiles releases by the U.S. General Services Administration.

The copper market has also been subject to several modeling efforts. Most notable, Fisher, Cootner and Bailey (1972) built a world copper model which was recognized as one of the first major econometric mineral modeling efforts. Their model divided the world copper market into the United States, where prices are administered by U.S. producers, and the rest of the world, where prices are determined by free market forces of demand and supply at the London Metal Exchange (LME). Since the LME price is free-market determined, it also played a role in determining the U.S. producer price in the long run, as well as providing a link between the two markets. Interregional trade between the United States and the rest of the free market world, thus depended on the differential between the two market prices and provided a further link between the two regions. The model was used to answer several then important policy questions: (1) How would possible cartel behavior on the part of major copper exporting countries affect the market; (2) What would be the effect on Chilean revenues and world prices of increases in Chilean exports; and (3) How would major new copper supplies affect the LME price?

Another major concern in explaining the wide and frequent price and quantity fluctuations in mineral markets has been the modeling of the disequilibrium as compared to the equilibrium characteristic of these markets. This can be most strongly seen in the role that stocks, either in the form of capital assets (including capacity and reserves) or inventories, play in mineral markets. CRA (1978) thus extended the supply sector of the Fisher, Cootner and Bailey (1972) copper model to include long run adjustments in exploration and discovery as well as subsequent mining capacity formation. This long run adjustment process was combined with a short-run inventory adjustment process in a distinctly disequilibrium form of copper model by Labys (1982b). Such an approach to modeling the copper market was suggested by Richard (1978) with his continuous time, differential equation approach to modeling the copper market.

CONTROLLED MARKET MODELS

While the above market model can be adapted to include the influence of market regulation, it remains a competitive model. This predicament can lead to serious consequences when modeling energy markets whose structure tends to be noncompetitive (Lord, 1991). That is, their structure may vary from complete control in the form of government monopoly to lesser degrees of noncompetitive behavior such as that of duopoly or oligopoly. The principal transformation that must come about in describing market behavior is to model controlled price determination from the point of view of the actions of individual market participants or of government policy, rather than of the workings of the market as a whole.

The several econometric approaches taken to model these noncompetitive market configurations are essentially similar. For example, the monopoly case involves one dominant (monopolist) producer and many (perfectly competitive) consumers. The single producer thus maximizes his own profits given the aggregate demand function for the commodity of interest and the supply response of the other firms in the industry.

The simplest case to envision is that of an oil cartel which sets prices P to maximize profits, but where the fringe S sets the quantity supplied where X is the quantity supplied by the cartel and D is the total market demand.

$$X_t = D_t - S_t \tag{4.5}$$

$$D_t = b_o - b_1 P_t + b_2 A_t \tag{4.6}$$

$$S_t = b_3 + b_4 P_t \tag{4.7}$$

Solving all three equations results in the following reduced form for the profit-maximizing price.

$$P_t = c_o + \frac{D_t}{b_1 - b_3} - \frac{S_t}{b_1 - b_4} \tag{4.8}$$

A variant of this form of model was constructed by Blitzer et al. (1975) to analyze the previous behavior of the OPEC cartel in the world petroleum market. Only now prices are assumed to be given and production of the cartel and fringe are determined. The model they postulate is as follows:

$$X_t = D_t - S_t \tag{4.9}$$

Prices can be determined from the relation

$$D_t = b_o - b_1 \, P_t + b_2 \, A_t \tag{4.10}$$

$$S_t = b_3 + b_4 \, P_t \tag{4.11}$$

$$P_t = \sum_{j=1}^{n} \alpha_j \, P_{t-j} \tag{4.12}$$

or they can be assumed to be an exogenous policy variable. One can also assume that maximum production $\overline{X}$ can be attained over and above normal production X.

$$\overline{X_t} \geq X_t \geq 0 \tag{4.13}$$

The criterion function of the cartel can be defined as

$$\pi_t = \sum_{t}^{n} g(F_t, \, t) + h(Y_t) \tag{4.14}$$

with $F_t = f(P_t, X_t)$

where:

D	=	Fuel market demand
S	=	Fuel supply from fringe
X	=	Fuel supply from cartel
P	=	Fuel prices
A	=	Income or activity level
N	=	Geological factors
K	=	Production capacity
π	=	Criterion function of cartel
F	=	Foreign reserves of cartel
Y	=	Indirect profits of cartel
θ, γ, α	=	Lag distribution parameters

Allowing for an appropriate time period of adjustment, market equilibrium is given by equation (4.9). Demand relations and fringe supply relations are formulated as in the competitive model. However, the authors interpret the supply relation (4.11) to be short run in nature since it assumes capacity to be fixed. They thus prefer to add a long run relation by first introducing capacity K into (4.11).

$$S_t = b_3 + b_4\,P_t + b_5\,K_t + b_6\,N_t \qquad (4.15)$$

Capacity further depends on fuel prices P after an appropriate gestation lag and on lagged demand, thus reflecting long-run investment decision making.

$$K_t = b_7 + b_8\,P_{t-\alpha} + b_9\,D_{t-r} \qquad (4.16)$$

This simple model is closed assuming a long-run distributed lag of prices P given by (4.12) where the weights α_j sum to unity and investors have a "memory" of n years when forming price expectations. This is the "so-called" non-competitive market model based on the assumption of exogenous prices. The weights α_j can be said to vary according to government policy formation and their range would depend on government pricing policies. To remove the price assumption, one could compute prices as a function of a set of output policies assigned to X. The constraint equation (4.13) completes the model by employing a vector of maximum production levels X in a given year.

Profit seeking or maximization is viewed through the policy criteria function (4.14). The value of π is assumed to depend on foreign reserves F gained and indirect profits Y. No actual maximization of the cartel's profits takes place in the form of a feedback control mechanism. Rather different pure-production policies are set and the criteria function evaluated over some time horizon to determine what might be an optimal policy or set of policies. Of course, the criteria function as well as the other relations would be specified more complexly in an actual model.

Among the development of econometric mineral models relating to monopolist market configurations, one of the few is that of Burrows (1971) who modeled cobalt producers as acting like an effective cartel. Unlike copper, tin, and many other mineral commodities, the market structure of cobalt is highly concentrated in production. One company, Union Miniere Haut Katanga (UMHK) has produced more than 60 percent of the world output, the rest being produced by various companies in Canada at 8 percent and in many other countries. Such a concentration on the supply side rightly warrants allowances for market imperfections in model specification. The general structure of the model is derived by treating UMHK as a price setter following profit-maximization principles (given the supply response of all other producers).

Regarding intermediate mineral market configuration, Pindyck (1978a) has applied his model to determine optimal price and quantity paths that would result from cartel behavior on the part of producer's organizations in the copper and bauxite markets. Other attempts to model intermediate market structures have not emphasized optimization but rather various forms of producer pricing systems. In an intermediate mineral market, the producers price is often modeled simply as a constant of markup over costs. A certain amount of material in such markets will be sold at open market transactions prices. In times of very "weak" markets (low demand at the list price relative to capacity), producers are apt to allocate sales at the list price. The prices of material sold by consumers and fringe producers on the open market will then include a substantial premium.

In modeling intermediate markets closer to monopoly, I have shown how the market equilibrium equations can be replaced with a list price or administered price equation in which price is modeled as a function of cost. (This procedure, of course, requires that accurate data be collected on the production costs of the dominant producer or producers.) Depending on the modeler's perception of the relative strengths of the monopoly group of producers and the fringe producers, prices may be modeled in one of two ways. If the monopoly group is relatively strong, the list price can be a function of cost and transactions price; this depends on a measure of excess capacity. If the monopoly group is relatively weak, the transactions price can be a function of demand and inventories, and the list price a function of the transactions price and a measure of cost pressures. Examples of such models include those constructed by CRA for nickel (1974), molybdenum (Burrows, 1972), aluminum (Burrows, 1972) and by Gupta (1982) for zinc.

It is important to realize that the character of intermediate markets can change over time for both cyclical reasons and for longer-term structural reasons, such as the entry of new producers. Thus, energy and mineral markets can be portrayed as a spectrum ranging from perfect competition to pure monopoly, with the markets for individual metals constantly shifting back and forth along the spectrum. Further comments on the specification of these kinds of models appears in the Chapter 6 on resource exhaustion models.

DEMAND MODELS

Certain econometric methods have been developed to explicitly analyze energy demand, although process and input-output methods have also been employed. There are three important reasons for modeling energy demand. The first is that demand forecasts can help to plan energy supplies, which are vital for the functioning of a modern economy. The second is that the expansion of these supplies usually requires many years. And the third is that investments in energy supply systems generally are highly capital intensive, on average, accounting for some 30 per-cent of gross investments in most countries. If demand forecasts are too low, energy shortages may develop whose costs are usually a large multiple of the volume of energy not supplied; but if forecasts are too high, large amounts of capital with high opportunity costs might be uselessly tied up for long periods of time. Among the econometric approaches employed for forecasting energy demands are: output demand models, price demand models, translog or functional form models, systems demand models, panel data models, and discrete choice models. Several reviews of these approaches appear in works by Griffin (1991), Phlips (1974), Boyd (1992), Donnely (1987), Theil and Clements (1987), Madlener (1996), Munasinghe (1990), and Sterner and Dahl (1992). Major areas of application include industrial demand, residential demand, and transport fuel demand.

The earlier energy demand models assumed that energy was tied to output, e.g. Schurr (1960). This relation was typically established using a Leontief production function in which output Q depended on value added VA(K,L), energy E, and materials M inputs. The technical coefficients γ_1, γ_2, and γ_3 measure the Leontief coefficient for the dollar value of any given input necessary to produce a dollar's worth of output.

$$Q = \min(\frac{VA(K,L)}{\gamma_1}, \frac{E}{\gamma_2}, \frac{M}{\gamma_3})$$
(4.17)

The value added aggregate could in turn involve a Cobb-Douglas or CES relationship among capital K and labor L. Taking GNP as a proxy for output, energy demand can be tied to GNP through a technically determined coefficient (δ) as follows.

$$E = \delta GNP$$
(4.18)

Thus energy demand forecasts depended entirely on GNP growth. Refinements of the methodology attempted to explain changes in γ_2, linking it to changes in the technical efficiencies governing autos, railroads, refrigerators, and so forth. Having determined aggregate energy consumption in this manner, researchers then proceeded to allocate BTUs among the four primary fuels, hydroelectric, coal, natural gas, and oil based on the supply of these fuels.

In this system, prices were largely irrelevant both in the determination of aggregate energy demand or the supply/demand functions for the individual fuels. Lacking a demand specification for individual fuels, price determination was handled either in an *ad hoc* manner or treated as exogenously determined. Natural gas prices and crude oil prices in the United States were administratively determined, obviating any need to model these prices. Because of the perceived unimportance of prices, the analysis tended to take on an engineering orientation with a focus on technical factors affecting both supply and demand. However, important studies such as that of Balestra and Nerlove (1966) revealed the importance not only of own prices but also of relative prices in demand determination.

The earlier forms of energy price-demand models were concerned with the residential, commercial, and/or industrial sectors and were single equation, long-run equilibrium demand models focusing on a single fuel. Such models were typically static and explained the demand D for a fuel as a function of its own price P, the price of competitive fuels PC, and industrial activity A or national income.

$$D = d(P,\ PC,\ A)$$
(4.19)

While the characteristics of the fuel burning equipment are reflected in the substitution of competing fuels, the adjustment in the capital stock to changes in fuel demand is assumed to be instantaneous.

Such models became widely applied in attempts to employ price and output elasticities for demand forecasting purposes. The most generally used model is the so-called double elasticity model in which the energy demand E (total or sectoral) is related to an indicator of activity or income A and to energy prices P

$$E = A^a P^b \qquad (4.20)$$

where a and b are respectively the output (or income) elasticity and the price elasticity.

In such a model, consumer reactions to price variations is assumed to be immediate, whereas in reality there is some inertia due to the facts that consumers will in a first step modify the use of their equipment and in a second step adapt or change their equipment. This is why a simple double-elasticity model is often adapted to distinguish between short-term (i.e. with constant equipment) and long-term (5-10 years) responses to price changes.

In this case, the short and long-term demand are integrated into a single equation or dynamic model in which the equipment is not considered explicitly but is accounted for by means of a lagged variable. Short and long-term price elasticities are a/1-c and b/1-c respectively. However, differences between short and long term elasticities can also be measured using the partial adjustment model. Let (4.19) now represent desired demand D^*.

$$D^* = d(P, PC, A) \qquad (4.21)$$

Partial adjustment assumes that the changes in actual demand D from the time period t-1 to t will adjust partially by λ to desired demand changes.

$$D_t - D_{t-1} = \lambda (D_t^* - D_{t-1}) \qquad (4.22)$$

By combining these two equations both short- and long-run demand responses can be differentiated.

Extending this model to a multi-equation formulation permits a short-run equation based on a fixed capital stock K to be separated from a long-run equation that explicitly includes the size and characteristics of that stock, such as in equation (4.17).

$$D = D^* = d(P, PC, A) \cdot U \cdot K \text{ A)} \tag{4.23}$$

and

$$K = K_{-1}(1 - \delta) + \Delta K \tag{4.24}$$

$$\Delta K = k(P, A, KC, PC) \tag{4.25}$$

where U is the capital utilization rate, δ is the rate of retirement, and KC includes capital costs and related stock efficiencies.

More recently it has been found useful to replace the fuel substitution effect based on the cross-price elasticity of PC with an interfuel substitution model that deals with competition from other fuels explicitly and in more detail. These models generally assume that the demand for any fuel cannot be adequately assessed without quantifying the price and sometimes the capital cost and non-price competition posed by other fuels (and where appropriate their respective fuel burning devices). These models treat the demand for energy such that capital K, labor L, energy E, and all other intermediate inputs M are seen as inputs to a production activity, when the output is defined as QO. The result is a production function of the following form known as KLEM.

$$QO = q(K, L, E, M) \tag{4.26}$$

This form has been developed as a translog function which possesses a flexible functional form.

The basic appeal of the translog and functional forms is that they relax the range of substitution possibilities in a production technology. For example, instead of requiring a constant elasticity or unitary elasticity of substitution among all inputs, it is possible to estimate KLEM models in which some inputs could be substitutes and others could be complements. These forms, usually but not necessarily, assume that production is characterized by constant returns to scale and that any technical change affecting K, L, E and M is Hicks-neutral. For given input prices and an output level, energy demand can then be determined simultaneously with the other inputs, assuming cost-minimizing behavior.

The actual determination of the fuel substitution effects is based on a form of disaggregation of energy demand E. That is, one can proceed sequentially to find the interfuel substitution responses based on a homogenous energy aggregate E that depends solely on fuel inputs F_i. For example, with four fuel inputs the energy aggregate would be

$$E = e(F_1, F_2, F_3, F_4) \qquad (4.27)$$

The determination of individual fuel usage occurs in this second step. This is due to the economic theory of separability, which permits individual fuel choice to occur independently of the choice of other inputs.

The econometric modeling of this approach has been largely through the translog share model, such as that developed by Berndt and Wood (1975). By treating the price of energy as a single argument rather than including the prices of the respective fuels, the model implicitly embodies the mentioned separability. Corresponding to the energy aggregate in equation (4.27) is the unit energy cost function for the energy aggregate.

$$P_E = \phi(P_{F1}, \ldots\ldots, P_{F4}) \qquad (4.28)$$

The sufficient conditions for this particular representation are of importance. Separability occurs if the ratio of the cost shares of any two fuels is independent of the prices outside the energy aggregate, such as labor or capital prices. In effect, the ratio of the fuel cost shares depends only on fuel prices. Linear homogeneity in input prices implies that the cost shares of fuels are independent of total expenditures on energy.

The system actually estimated is based on the following translog cost function

$$\text{Ln } P_E = +1/2 \Sigma \beta_{ii} \cdot \text{Ln} P_{Fi} \cdot \text{Ln} P_{Fj} + \alpha_0 + \alpha_i \text{ Ln} P_{Fi} \qquad (4.29)$$

where $\beta_{ij} = \beta_{ji}$. Because of the multicollinearity problems associated with the direct estimation of (4.29), researchers have utilized Shepherd's lemma to derive a set of cost share S_i equations:

$$S_i \alpha_i + \sum_j \beta_{ij} \text{Ln } P_j \qquad (4.30)$$

Shepherd's lemma requires that firms minimize costs and treat input prices as given. Partial differentiation of this equation yields a series of fuel cost shares S_i for the four fuels.

$$S_i = \frac{P_{Fi}F_i}{P_F E} = \frac{Ln\, P_F}{Ln\, P_{Fi}} = a_i + \sum_{i=1}^{4}\beta_{ij}\, Ln\, P_{Fj} \qquad (4.31)$$

Given that $\Sigma\, S_i = 1$ and imposing the following restriction, the final solution for the shares S_i requires estimating only three regression equations of the form (4.29).

$$\sum_i \alpha_i = 1, \sum_j \beta_{ij} = 0, \quad and \quad \beta_{ji} = \beta_{ji}\,(all\ i, j, i \neq j) \qquad (4.32)$$

Regarding the need to determine the substitution possibilities among the fuels, this can be accomplished by using the Allen elasticities of substitution (AES) defined for the translog as
or

$$\sigma_{ij} = \frac{\beta_{ij} + S_i\, S_j}{S_i\, S_j} \qquad (4.33)$$

$$\sigma_{ij} = \frac{\beta_{ij} + S_i^2\, S_j}{S_i^2} \qquad (4.34)$$

The AES are not constrained to be constant but may vary with the values of the cost shares. Indeed, Berndt and Wood's (1975) finding of energy-capital complementarity proved to be a catalyst, triggering both controversy and widespread interest in the translog function.

Other studies have used this approach in an attempt to evaluate and to measure various elasticities of substitution as well as related demand and supply price elasticities. Several surveys of energy demand elasticities have been made by Dahl (1993) and Dahl and Sterner (1991). Reviews of metal demand, cross-price and substitution elasticities are available from Hazila and Kopp (1984) and Fanyu (1996). The latter study additionally has used this approach to estimate elasticities for aluminum, copper, lead, nickel, tin and zinc. Comparisons are made with elasticities derived from the partial adjustment approach to demonstrate a greater consistency in the magnitude and sign of these elasticities using the KLEM or the flexible functional form model.

Griffin (1991) points to some further developments of the KLEM approach. By relaxing the assumption of a homogeneous unit cost function, researchers found the translog useful in testing for non-homothetic, non-constant returns to scale technology. In a study of cost functions for electric utility plants, Christensen and Greene (1976), appended

$$\delta_o \, Ln \; Q + 1/2 \; \delta_1 (\, Ln \; Q \,)^2 + \Sigma \gamma_i \, Ln \; Q \, Ln \; P_i$$

to equation (4.27) and $\gamma_i \, Ln \, Q$ to equation (4.28) producing a non-homogeneous, non-homothetic cost function. Tests for scale economies then depend on the elasticity of cost with respect to output (ε_{CQ}) as follows:

$$\varepsilon_{cq} = \delta_o + \delta_1 LnQ \qquad (4.35)$$

Another development was that the traditional time trend T used to reflect a constant rate of Hicks' neutral technical change could be generalized by introducing $\Phi_o T + 1/2 \, \phi_1 (t)^2 + \Sigma \psi_j \, T \ln P_j$ into equation (4.29), which in turn resulted in the addition of ψ_j and T terms into the share equations in (4.30). Consequently, technical change T became measured as follows:

$$T = \Phi_0 + \Phi'_1 + \Sigma \psi_j \, Ln \, P_j \qquad (4.36)$$

This specification allowed technical change to be both factor saving or using (if $\psi_j \neq 0$) and proceed at differential rates over time (if $\Phi_1 \neq 0$). Jorgenson and Wilcoxen (1991) further suggested that technical change tended to be energy using so that the productivity decline in the 1970s was attributed in part to higher energy prices.

Despite the numerous advantages of the translog function enumerated above, practitioners attempting to use translog models for forecasting and simulation analysis noted two serious limitations. First, the concavity conditions that assure the factor demand function to be negatively sloped may be satisfied at the sample mean, but may misbehave over other price regions. Unfortunately, concavity is satisfied globally only when the translog reduces to the special case of the Cobb-Douglas function. To this end, Diewert and Wales (1987) have developed a constrained estimation procedure that imposes concavity over specified ranges of price variation.

Second, the translog model implies static expectations and instantaneous adjustment to long-run equilibrium. In effect, the firm is assumed to move instantaneously from one long-run cost minimizing equilibrium to another as relative factor prices change. In energy applications, where long lags are known to exist because of the long-lived nature of the capital stock, this characteristic is particularly troublesome. In an attempt to provide a unified theoretical framework capable of describing the short-run equilibrium and the adjustment path to the long-run equilibrium, Berndt, Fuss, and Waverman (1980) developed a model in which capital was costly to adjust. Besides yielding implausibly rapid adjustment to long-run equilibrium, their model relied essentially upon static expectations. Subsequently, Pindyck (1980) incorporated rational expectations in a variable cost function with a capital stock equation that adjusts optimally over time. Because of difficulties in solving stochastic optimal control problems, the dynamic adjustment path can only be simulated for deterministic price scenarios, limiting its use as a predictive device.

Industrial Energy Demand

Applications of the translog form of this model can be seen in the work of Berndt and Wood (1975) who analyzed interfactor substitution between energy, labor and capital in industrial energy demand. Their model concentrates on the derived demand for energy based on a fixed output. Emphasis is placed on the determinants of outputs, substitution possibilities among inputs allowed by the production technology, and the relative prices for all inputs. Extensions of this model dynamically can be seen in Hartman (1979) who analyzed short and long-run price elasticities of energy demand. While the above research concentrates on U.S. energy demand, application at the international level can be found in the work of Griffin and Gregory (1976) and of Pindyck (1979). One final application by Carson et al. (1981) combined the international demand model with a petroleum supply model to yield a demand and interfuel substitution analysis of the world oil market.

Regarding the application of the demand model in a regulated or noncompetitive context, only very limited research has taken place. One example of an attempt to include market imperfections in interfuel substitution models can be seen in the research of Fuss (1980). The model assumption normally made is that of infinitely elastic supply curves for the energy inputs. However, Fuss demonstrates that the model can still be used in a supply constrained market if the assumption is added that the producer continues to optimize subject to the constrained opportunities and the production technology. In this case market prices for the constrained inputs are replaced by shadow prices. His application deals with Canadian gas and fuel oil constraints and their impacts.

Wood and Spierer (1984) have also shown how natural gas regulatory pricing and supply constraints can be modeled. In this approach, disequilibrium effects due to regulation or supply constraints of one or more inputs in Swiss industrial gas use are explicitly modeled by treating the regulated input as a quasi-fixed input. Systems of fuel demand equations are derived and estimated from a variable cost function which is conditional on the quantities of natural gas and outputs. An important feature of this approach is that full equilibrium values for natural gas can be evaluated and used in calculating full-equilibrium price elasticities. For a regulated input, this approach eliminates biases in short-run elasticities for variable inputs due to disequilibrium caused by regulation. It also provides a means to test for the effects of regulation by comparing actual and optimal (derived from the variable cost function) values of the regulated input.

One problem in employing the translog approach to evaluate industrial energy demand is that the price elasticities often seemed not to approximate reality. For example, Hudson and Jorgenson (1974) found that the implied price elasticity for energy was perverse in the manufacturing sector. Furthermore, the own and cross price elasticities among fuels were close to zero. Also, using aggregate time series data, Berndt and Wood (1975) interpreted their finding of energy-capital complementarity as a long-run result rather than as a short-run anomaly due to the fixity of the energy efficiency of the capital stock in the short run. As surveyed by Bohi (1989), time series studies of gasoline and electricity demand characteristically found price inelastic demand responses as well.

In contrast, researchers utilizing cross section and panel data sets tended to report much larger price elasticities, reflecting the fact that interregional price differences were both quite substantial and persistent so that the "between" region demand variation tended to reflect long-run responses. Particularly for energy demand, the adjustment to the long run can take ten or more years because of the long-lived nature of the capital stock to which energy consumption is tied. Consequently, one would expect intercountry cross section data to elicit long-run responses while time series data would reflect incomplete adjustment.

This result has lead to significant increases in both the quality and number of panel data studies. A striking attribute of these studies was the application of a battery of tests and a variety of feasible GLS estimators. Tests for one or two-way error components, poolability, and specification error became commonplace. Researchers increasingly perceived that intercountry and interregional price and income variation provided valuable information not usually available from simple aggregate time series data. Baltagi and Griffin (1988) in their electric utilities study have proposed a method of estimating a purely general index of technical change which does not rely on linear or quadratic functions of time. They accomplish this by applying non-linear estimation techniques to a panel data set of firms within the context of a translog cost function.

Residential Energy Demand

Another application of the translog model has been to explain household or residential energy demand, along with several other methodologies. The formulation of a typical translog residential energy demand model is based more on consumption theory, i.e. see Madlener (1996) and Munasinghe (1990). The direct utility function of a consumer which indicates the intrinsic value derived from the consumption of various goods may be written:

$$U = U(Q_1, Q_2, ... Q_n, Z) \qquad (4.37)$$

where Q_i represents the level of consumption of good i in a given time period and Z is a set of parameters representing consumer tastes and other factors. The set of prices, P_1, P_2, ...P_n, for these n consumer goods, and consumer income I, define the budget constraint:

$$\Sigma P_i Q_i \leq I \qquad (4.38)$$

Maximization of the consumer's utility U subject to the budget constraint yields the set of Marshalliam demand functions for each of the goods consumed by the household

$$Q_i = Q_i(P_1, P_2, ... P_n; I; Z) \ for \ i = 1 \ to \ n \qquad (4.39)$$

Consider the demand function for a particular fuel (e.g., gas). Then equation (4.39) may be written in the simplified form:
where the subscript g denotes gas, while subscripts e and o indicate the

$$Q_g = Q_g(P_g, P_e, P_o, P, I, Z) \qquad (4.40)$$

where the subscript g denotes gas, while subscripts e and o indicate the substitute forms of energy, electricity and oil (possibly for cooking), and P is an average price index representing all other goods.

Next, assuming that demand is homogenous of degree one in prices and income, one can write:

$$Q_g = Q_g (P_g/P, \ P_e/P, P_o/P, I/P, Z) \qquad (4.41)$$

Thus, starting from consumer preference theory, a demand function can be derived for a given fuel which depends on its own price, the prices of substitutes and income, all in real terms. The effects of other factors Z such as quality of supply, shifts in tastes, and so on can also be added.

The final specification of an equation such as (4.41) could vary widely (Taylor 1977, Pindyck 1979). Q_g could be household consumption or per capita consumption; the demand function could be linear or linear in the logarithms of the variables or in the transcendental logarithmic form and could include lagged variables; and Z could include supply side constraints such as access to supply, and so on.

These models have been used mainly to measure the effects of changes in income and prices on residential energy demand, principally in the form of price and income elasticities. Such applications had earlier been surveyed by Bohi (1981), Hartman (1978, 1979), Mitchell et al. (1986), and Taylor (1975). Most recently Madlener (1996) has classified residential energy demand models according to: (1) log-linear models, (2) translog models (3) qualitative choice models, (4) pooled cross-section/cross-country models, and (5) time series models. Among these, the log-linear form reflects flexible functional forms, including the above translog and the generalized Leontief functional form.

Regarding residential substitution and fuel choice, qualitative choice models explain this behavior using logit, probit, and tobit approaches. Since Fisher and Kaysen's (1962) analysis of appliance stocks and electricity consumption, economists have recognized that the demand for a given fuel F depends first on the stock of energy consuming equipment S, then on the efficiency e with which the fuel is utilized, and finally on the utilization rate U of the capital equipment:

$$F = \frac{S}{e} \cdot U \qquad (4.42)$$

According to Griffin (1991), for any given capital stock, the efficiency of fuel conversion e is fixed by technical considerations, so that in the short run, the relevant decision variable affecting fuel consumption is the utilization of the capital stock. Interfuel substitution effects occur only in the long run with the introduction of new capital. For example, the choice of electric, natural gas, or heating oil for home heating is dependent on the choice of capital equipment and once installed, households can only vary their utilization of the heating type. Despite this intuitively appealing framework, most residential demand models made little use of this approach, partially because of inadequacies in capital stock data and partially because of the conceptual difficulties of discrete choice models.

McFadden and others however, have met this challenge by modeling appliance choice as a discrete choice within the utility maximization paradigm. By nature, the choice of an appliance type such as natural gas home heating instead of electric or oil heating is a discrete choice. Rather than focusing on the amount of a good consumed, one must first focus on the probability of choosing a particular appliance type. Because of the bounded nature of the probabilities, simple regression models are no longer adequate. Consequently, the multinomial logit function has been employed. This model states that the probability of selecting appliance portfolio i from among n portfolios depends on the following relation

$$P(i) = \frac{e^{v_i}}{\sum_{j}^{n} e^{v_j}}$$
(4.43)

where v_i and v_j are functions determining the level of utility associated with any respective portfolio. These functions are specified from an underlying indirect utility function in which the discrete choice follows from a random utility maximization process whereby consumers attempt to maximize the utility from alternative appliance portfolios. In addition, the appliance choice decision and the utilization decision are modeled simultaneously, avoiding the bias due to correlation of an explanatory variable and the disturbance term in the utilization equation.

The estimation of this model is based on household survey data, which contain detailed information on economic and demographic characteristics of the households as well as their stock of various energy using appliances. According to Griffin (1991), one difficulty with the use of a single cross-section is that the appliance choice decision is often made at the time of home construction, which does not generally coincide with a sample period. Therefore, the prices, appliance choice, and utilization rates do not reflect long-run maximizing behavior, but instead reflect maximization given the appliance stock. Thus, the true long-run price elasticities may be much greater. As longer panel data sets for individual households become available, it will be possible to separate data points where the appliance choice and utilization decision are truly simultaneous and the more usual situation of a utilization decision conditional upon a given appliance stock.

These various methodologies thus provide insights into how residential energy demand can be met, given the choices that have to be made in selecting fuels which also support environmental quality.

Transport Fuel Demand

A number of models have been developed to explain fuel demand for vehicles and transport. In particular, Sterner and Dahl (1992), Dahl and Sterner (1991), and Sterner (1990) have analyzed the various studies dealing with the demand for gasoline. The simplest such models begin with the assumption that the fuel utility function embodies motor fuel demand G and

$$U(G, A) + h(I - PG - PCA) \qquad (4.44)$$

an aggregate of other goods A. Consumers thus maximize utility U(G,A) subject to a budget constraint PG + PCA ≠ I, where P is the price of gasoline, PC is the the price of other goods, and I is income. Gasoline demand thus depends on gasoline prices, other prices and incomes.

$$G = f(P, PC, I) \qquad (4.45)$$

or, assuming log-linearity,

$$\text{Ln } G = a_0 + a_1 \text{ Ln } P + a_2 \text{ Ln } I + \alpha_3 LnPC + \varepsilon \qquad (4.46)$$

The most crucial problem with this model is that there may not be time for demand to adjust to changes in price and income within a given period. As mentioned above, such a problem can be handled with a partial adjustment mechanism involving time t where the desired consumption of gasoline G* adapts accordingly to the fraction s given by (4.48).

$$G_t^* = a_0 + a_1 P_t + a_2 I_t \qquad (4.47)$$

$$G_t - G_{t-1} = s(G_t^* - G_{t-1}) \qquad (4.48)$$

$$G_t = sa_0 + s\alpha_1 P_t + s a_2 Y_t + (1-s)G_{t-1} + e_t \qquad (4.49)$$

Combining (4.47) and (4.48) and rearranging gives the familiar result.
This equation thus yields both the short and long run elasticities.

$$G_t = b_0 + b_1 P_t + b_2 I_t + b_3 G_{t-1} + e_t \qquad (4.50)$$

Gasoline demand can also be modeled by including the capital stock as was shown earlier. The simplest case assumes that consumers purchase gasoline G and automobile services S to produce transport miles. Aggregate formulations of this equation in either a static or dynamic framework have been used in a few studies. In practice, however, the most common approach has been to assume, in the short run, that the stock of autos K is fixed.

$$G_t = a_0 + a_1 P_t + a_2 I_t + a_3 K_t \qquad (4.51)$$

Finally, as in the case of industrial and residential modeling, Baltagi and Griffin (1983) argue in favor of using pooled cross-section time series data to improve the elasticity estimates. However, they disregard the adapting of the number of vehicles to the price of gasoline and instead model gasoline per vehicle G/A as the utilization U (M/A) divided by the efficiency E (M/A). Utilization can be modeled by a formula analogous to what is called the simple vehicle model (including income per capita, gasoline price and the stock of vehicles per capita). Efficiency subsequently can be modeled as a function of per capita income and gasoline price but with distributed lags, i.e. see Archibald and Gillingham (1980).

SUPPLY MODELS

These models are concerned with the supply of non-renewable resources. However, the construction of models based purely on historical data has had some difficulty because of the problem of interrelating economic concepts of Hotelling rents with influences involving technological changes and geological conditions. This area has seldom been traced as a separate class of mineral or energy models, but it does embody distinct contributions stemming from the work of Adelman (1983) and Fisher (1964) on crude oil, Erikson and Spahn (1971) and MacAvoy and Pindyck (1973) on natural gas, Harris (1984) on minerals, and Zimmerman (1977) on coal. The methods applied range from a combination of econometrics and geostatistics to dynamic optimization.

Most oil supply models represent a combination of economic supply and exploration determinants. A major exception are the Hotelling-based models such as that of Cairns (1986) dealing with regions or Pindyck (1978) dealing with depletion, as described in the section on resource exhaustion models. Padilla (1992) has emphasized the difficulties of modeling oil exploration. The underlying conditions are very heterogeneous, since they reflect a multitude of distinct historical, economic, legal, political, geological, cultural and other factors. Clearly the type and relative importance of the factors depend on the particular situation of a geological region. However, the results of econometric analyses also show that non-market factors play a very important role in explaining exploration behavior. In some cases, their importance is such that even the price of oil has a negligible influence. Siddayao (1980), for example, finds that prices do not have a significant statistical influence on exploration drilling in South-East Asian countries. And Broadman (1985) concludes that prices go some way in explaining the extent of exploration effort in oil producing developing countries but that the correlation between exploration effort and price is negative in the case of non oil-producing countries.

Most theoretical models approach the problem of oil supply, including exploration, either as one of constrained intertemporal maximization (i.e., Pakravan, 1977; Peterson, 1978) or as a problem of stochastic optimization under uncertainty, i.e. Gilbert (1979); Pindyck (1980); Arrow and Chang (1982). In a competitive market, exploration can be said to be a function of the scarcity value of the rent underground. The latter can be measured using prices (e.g. Gaudet and Hang (1986)) or by marginal costs e.g. Devarajan and Fisher (1982). The models which have been based on these relationships can be divided into two groups. The first group is based on concepts of inter-temporal maximization, i.e. see Epple (1975, 1985), Cox and Wright (1976), Nielssen and Nystand, (1986). The second group employs an expected profitability function for exploration based on exploration-related investment decisions. The expected profitability function can be estimated in several ways. For example Uhler and Eglington (1983), Scarfe and Rilkoff (1984), Ryan and Livernois (1985), and Bing (1987) use the expected profits to explain exploration E measured by land expenditures and also by the depth.

$$E_t = A_0 + A_1 E_{t-1} + A_2 Q_t + A_3 \Pi^{oil} + 4_4 \Pi^{gas} + a_5 DU \qquad (4.52)$$

Here Q is the output in million cubic meters of oil equivalent; Π^{oil} and Π^{gas} are the expected half-cycle profits on oil and gas respectively; and DU is a dummy variable introduced to cover the effect of energy policy.

An alternative econometric approach adapted by Moroney and Bremmer (1987) and Desbarats (1989) use "netbacks". The net price for the producer or "netback" is defined as the wellhead price of oil (or gas) less production costs, taxes and royalties. In the absence of a credible explicit measure of profits in the oil industry, the (observable) producer's net price is a good approximation of companies' cash flow available for future investment. It also provides a proxy cost for the acquisition of reserves. This indicator has the advantage that it does not require hypotheses concerning the operators' (unobservable) outlook regarding the future of prices, costs, taxation etc. Moroney and Bremmer, for example, take netbacks as a starting point for the construction of a variable which they call "profit", which is intended to reflect exploration profitability. Thus, the producer's net prices for oil and gas are calculated first before going on to estimate the production profile of the discovery made in year i. Multiplying the netback for the year of the discovery by the volume of hydrocarbons extracted each year provides an estimate of the annual revenue over the life of the discovery. Future revenues are then summed and discounted in order to obtain a discounted net operating profit per barrel of oil or million cubic feet of gas. In order to take geological factors into account, the success rate and discovery rates (volume of oil and gas discovered per successful exploration well) are introduced into the explanatory variable "profit".

The resulting equation they estimate gives exploration drilling N as a function of lagged drilling and lagged expected profits Π, where expected profits Π are defined by

$$N_t = a_0 + a_1 N_{t-1} = a_2 \Pi_{t-1} \qquad (4.53)$$

$$\Pi_t = P_t^{oil}\left[\frac{S^{oil}}{N}\right]_t\left[\frac{R^{oil}}{S^{oil}}\right]_t + P_t^{gas}\left[\frac{S^{oil}+S^{gas}}{N}\right]_t\left[\frac{R^{gas}}{S^{oil}+S^{gas}}\right]_t - DC_t \quad (4.54)$$

Here P is the discounted net price, S represents successful oil (gas) wells, N defines total exploratory wells, R is the volume of oil and gas discovered, and DC is the average after-tax drilling cost per well. Other econometric approaches use the value of undiscovered deposits, or an index of average deflated revenue.

Finally, oil supply models also have been constructed to maximize complex objective functions which also reflect the above price and cost influences. Examples include the U.S. Department of Energy (1978) model and the Gas Research Institute oil model documented by Woods and Vidas (1983). In the former case, the DOE model calculates the quantity of economically exploitable resources in each of twelve oil regions in the United States. It then generates regional curves for "latent" or "desirable" demand for exploration drilling at each price level. The relation between latent exploration demand and price is unique for each region. Total demand is thus the sum of the regional latent demands and declines with a reduction in the quality of discovered reservoirs and cost increases. The model also considers total desirable drilling subject to several constraints such as availability, capacity and the economic life of drilling equipment in order to obtain the total quantity of "realizable" exploration drilling.

In the case of coal, the specification of supply functions is directed more towards an expression for average cost which varies with levels of output. The estimation of realistic or accurate supply curves is made more difficult because of the peculiar properties of coal, i.e. Zimmerman (1977, 1981). While data about coal seams can be abundant, a wide range of geostatistical, geomechanical and geochemical characteristics can profoundly affect the costs of mining. These include the size of the deposit, the thickness of the coal seam, the depth of the rock and soil overlying the seam (the overburden), the amount of tectonic disturbance, the angle from horizontal in which the seam occurs, the friability of the overburden, and the rate of inflow of water and methane into the mining section. Also important is the heterogeneity of coal grades or quality, its bulkiness, and the impurities which must be removed.

Including all of these factors in a coal supply function is difficult. The actual process of constructing a supply curve involves three steps: (1) the development of relationships between the physical conditions of mining and costs; (2) defining that portion of the coal endowment that can be developed and of potential economic interest over the period covered by the analysis, and (3) using the cost relationships developed in step (1) to transform the geological data into a potential supply curve. These models tend to be long run, emphasizing the relationship between costs or productivity and geological conditions such as the above, in order to measure changes that may be expected in costs, as these conditions deteriorate or depletion emerges. Reviews of coal supply models are numerous and, for example, include Gordon (1979), Energy Modeling Forum (1978), Goldman and Gruhl (1980), Steenblick (1985), Price (1984) and Wood and Mason (1982).

An econometric approach, developed initially by Zimmerman (1977) and refined by Barrett (1982), instead of estimating costs directly from physical conditions, defines the relationship between the productivity of the relevant producing units comprising a mine and the mine's in situ seam characteristics and output. In deep mining, a production unit or mining section is defined as a mining machine and its labor force. The corresponding measure in the surface mining model is the capacity of the overburden-removing equipment: the maximum usefulness factor of draglines (defined as the product of the volumetric capacity of the dragline's bucket and the length of its dumping reach) for area, or strip mining, and the shovel bucket capacity for open-pit mines. Zimmerman (1977) accordingly defines productivity as

$$q = Q \,/\, S = A\,Th^{\alpha}\,S^{P}\,Op^{\psi}\,E \qquad (4.55)$$

where q is the average mining section productivity; Q is the total annual output of a mine; S is the number of mining sections operating at the mine; Th is the coal seam thickness; Op is the number of openings (or shafts) to the mine; and E is a random disturbance term for the unobservable geotechnical characteristics.

Using this equation as a base, Zimmerman also estimated a long-run cost function. He showed that necessary expenditures on labor, operating supplies and capital could be approximated as a function of the required number of mining sections, using engineering estimates of expenditure categories for hypothetical mines. On the assumption that, in the long run, the marginal cost MC of producing coal will equal the minimum average cost AC of a mine, he then established the minimum efficient scale Q* of a mine in order to evaluate the marginal and average per-ton cost at that point. (Somewhat different means are used to determine Q* for underground compared to surface mines.)

$$MC = AC^{*} = K_{m}\,/\,Th^{\alpha}\,E \qquad (4.56)$$

where K_{m} is a constant corresponding to a particular mining technology. Steenblick (1992) has criticized the large unexplained residual found in the productivity equation.

Zimmerman (1983, p. 308) concluded that "if only the seam thickness for a given mine is known, a 90 percent confidence interval includes productivity levels almost seven times greater than those predicted by" the equation. To this end, Barrett (1982) has shown that considerable improvements can be obtained by performing the analysis on a smaller, regional basis, using empirical data from mines specific to that region and by including alternative technologies. Whereas Zimmerman assumed only continuous mining technology (whereby coal is ripped from the coal face and loaded onto a conveyor belt in one continuous operation), Barrett looked at three different technologies. The results suggest that the substitution of alternative technologies can off-set the long-run effects of depletion. When such substitution is lacking, Fettweis (1983) has shown that mining costs can be modeled as an increasing function of deteriorating geological conditions.

Concerning engineering cost or optimization approaches, Labys and Yang (1980) in their modeling of coal supply have shown how cost differences can affect coal spatial distribution patterns. Here cost elements such as initial capital, deferred capital, and certain elements of annual operating costs are estimated individually on the basis of relationships derived from engineering-cost models of representative mines. These model mines are hypothetical constructs, distinguished by size, mining method (e.g., surface or deep), and seam conditions. Rules are then developed to vary costs with variations in mining conditions from those assumed for the base-case model mines; and output is assumed to adjust optimally so that average costs are minimized. See also the Coal and Electric Utilities Model (e.g. ICF, 1977) and the CRA (1982) model of coal production, transportation and allocation.

ENERGY CONSERVATION MODELS

These models which constitute an extension of the KLEM model have been used to measure how energy saving technologies might impact on the related production factors in energy supply and ultimately on energy savings, i.e. see Jacques, et.al (1988). Lesourd (1984) defines this modeling approach by including technology T as an explicit variable rather than being embodied in the functional form

$$Q = f (L, \ K, \ E, \ T) \tag{4.57}$$

Assuming that

$$v = \left(\frac{L}{E}, \frac{K}{E} \right) \ and \ h(v) = h\left(\frac{L}{E}, \frac{K}{E}, T \right)$$

energy efficiency can be defined at constant production Q^0 by:

$$\frac{Q^0}{E} = h(v_L, v_K, T) \tag{4.58}$$

For a simple interpretation of this model, consider the case where f (L, K, E, T) is homogeneous of degree 1; then h (LV) will be formally identical to f (K,L, E, T) and h (v_L, v_K, T) will be formally identical to f (L, K, E, T). As a result, the efficiency of energy given by (4.58) can be interpreted as a function of *three groups* of variables: (1) labour intensities v_L, (2) capital goods intensities v_K, and (3) variables describing the state of technological knowledge, or of scientific and technological progress, denoted by T. In this case then, changes in energy conservation or savings can be interpreted as varying with the following three periods in which input variables can be adjusted from being fixed to becoming variable.

The Short Term. The capital intensities v_K and the state of technological knowledge may be considered fixed variables, and energy savings may only be achieved by acting on labour intensities. It is therefore likely that factors such as management or labor, personnel training (or specific human capital), may be thought of as exhibiting increasing returns, and thus significantly increasing energy savings. It has been shown empirically by Smith (1976) and Lesourd and Consonni (1984), that such energy savings can be achieved without capital investment, and only by energy management actions such as energy accounting and control systems, auditing, organizations, personnel training, etc. Others also argue for multiple levels of energy management, among which some are short-term, including basic system design and the interpretation of the results of measurement and design.

The Medium Term. A second level of possible energy savings may be identified when the capital intensities v_K are considered as variable, in addition to the labout intensities v_L but under a given fixed state of technological knowledge. Then, energy savings amount to overhauling some process, shifting to a modification of that process available under the given state of technological knowledge. In short, significant energy savings at this level may be achieved by investing in components of capital exhibiting significant substitutability for energy, by anti-energetic investment (energy-saving investment), or by opposition to pro-energetic investment (i.e. in capital goods, that, by nature, consume energy, such as engines, ovens, etc.). Here, the expressions "anti-energetic" and "pro-energetic" capital are due originally to Le Goff (1979) and have also been used by Lesourd (1984). It has been shown empirically by Smith (1976) and Lesourd and Gousty (1981), that by cumulating the above energy saving actions, managers can often achieve energy efficiency gains of more than one-third. Some energy-saving investments such as investments in thermal insolation, in heat exchangers or in more process modifications all fall into this category.

The Long Term. A third level of energy savings may be identified when all three groups of variables in (4.57) (labour, capital and technological knowledge inputs) are allowed to vary. This level of savings corresponds to complete overhaul of a given process, shifting to a new technology not previously available.

This modeling approach has been stated to be valid for the more energy-intensive industries, mainly process industries, where energy is embodied in physiochemical or physical processes, and constrains the production function. It is likely that energy is not such a limiting factor in less energy-intensive activities such as, for instance, some agricultural and service activities (for which the direct energy content, is small). Hence, this approach broadens some usual neoclassical convexity hypotheses and gives a theoretical basis for the above translog KLEM function which can satisfy assumptions that are not met by the Cobb-Douglas function.

5 SPATIAL EQUILIBRIUM AND PROGRAMMING MODELS

While a common element of the above models is examining market behavior over time, another class of economic models is concerned with models describing commodity process at one point in time or representing commodity transfers over space. Such a modeling approach has been applied to both mineral and energy industries and derives from classical activity analysis or mathematical programming models. The methodology of linear and nonlinear programming and numerous practical applications are described by Dantzig (1963), Wagner (1969) and more recently Brooke et al. (1988). These kinds of models have been applied to mineral and energy markets in the form of linear or transportation programming, process programming, quadratic programming, mixed integer programming, linear complementarily programming, and variational inequalities. Further analysis of this approach can be found in Labys, Takayama and Uri (1989) and Labys and Yang (1991, 1997).

LINEAR AND TRANSPORTATION PROGRAMMING

Of the several methodologies available for solving mineral or energy spatial allocation problems, the Koopmans-Hitchcock transportation cost minimization model has received the widest application. It normally consists of three components: (1) a set of demand points or observations and a set of supply points, (2) the distribution of activities over space, and (3) the spatial equilibrium conditions. This model is typically stated in the form of an objective function C describing total costs subject to a set of constraints

$$Minimize \quad C = \sum_{i=1}^{n} \sum_{j=1}^{n} T_{ij} Q_{ij} \qquad (5.1)$$

subject to

$$D_i \leq \sum_j^n Q_{ij} \quad \text{for all } i \tag{5.2}$$

$$S_j \geq \sum_i^n Q_{ij} \quad \text{for all } j \tag{5.3}$$

$$Q_{ij} \geq 0 \quad \text{for all } i, j \tag{5.4}$$

where T_{ij} is the given transportation costs of shipping a commodity between region i and region j; D_i is defined as commodity demand in region i; S_j is commodity supply in region j; and Q_{ij} is the quantity shipped between region i and region j.

Transportation costs are minimized by allowing mineral and energy commodities to transfer until demand equals supply in every spatially separate market. The cost minimization process is established by the objective function (5.1). The constraint relations (5.2) and (5.3) reflect the conditions that regional consumption cannot exceed total shipments to a region, and total shipments from a region cannot exceed the total quantity available for shipment. Relation (5.4) precludes negative shipments.

The dual of the transportation problem identifies the set of mineral or energy prices at destinations and unit royalty payments that maximize total revenue less unit royalties. In the case of the simpler LP formulation, the dual determines a set of delivered prices and unit royalties (or economic rents) which is consistent with an efficient solution and perfectly competitive pricing. The objective function for the dual problem consists of total revenue R, which equals prices P_i multiplied by demand D_i and marginal input returns r_j (net of unit royalties) multiplied by production capacities S_j

$$Maximize \quad R = \sum_{i=1}^n P_i D_i - \sum_{j=1}^m r_j S_j \tag{5.5}$$

subject to

$$P_i - r_j \leq T_{ij} \tag{5.6}$$

$$P_{ii} \, r_j \geq 0 \quad \text{for all } i \text{ and } j \tag{5.7}$$

The profit condition of the price system is given by (5.6) and the set of prices and royalties which satisfies equation (5.8) is a solution for the price system.

$$P_j - r_i \leq T_{ij} \tag{5.8}$$

The relationship between the primary and dual programs of the transportation problem is that they are symmetric and that the minimum cost for the delivery system is equal to the maximum revenue for the price system. The dual problem establishes the shadow delivered prices in all markets. The shadow values determined in each solution indicate the cost savings to the economy of an extra unit of production, or of a unit relaxation in transportation mode constraints.

Early applications of linear programming to mineral industries include Copithorne's (1973) model for nickel which had a strong regional configuration. Kovisars (1975, 1976) extended the modeling application by including processing stages as well as temporal considerations in the cases of copper and zinc. These included mining, concentrating, scrap recycling, smelting, and refining. A temporal dimension was added by making successive adjustments to the demand, cost, and capacity variables based on the model solution at previous stages.

More recent applications have incorporated econometric equations, particularly on the demand side, and have emphasized dynamic model solutions. For example, Hibbard et al. (1979) constructed such a model for long range forecasts of important aluminum industry variables. Their model simulated the overall supply and demand balance of the major regions that constitute the world aluminum industry. The supply component is a regional and process-oriented, time-dynamic formulation of the flows of materials through mining, refining, smelting, scrap recycling, fabrication, and distribution of final products to end-use sectors. Supplies as well as demands in other regions of the world are determined using econometric relations. Assuming that the industry operates in a purely competitive environment, market prices of aluminum are determined by the intersection of marginal supply curves with demand curves. Hibbard et al. (1979) maximized the sum of the consumers and producers surplus to obtain a partial equilibrium analysis of supply and demand over time. Hibbard et al. (1980) also provided a similar modeling effort applied to the U.S. copper industry. Known as MIDAS-II, it features extensive disaggregation to include production characteristics of individual mine, smelter, refinery, and electrowinning units in the United States.

An additional application of linear programming to mineral commodities involves the lead and zinc model constructed by Dammert and Chrabra (1987) which determines the supply functions in markets where metals are mined as co-products. The programming sub-model determines what the minimum lead and zinc prices should be in order to cover the costs of production for the highest cost projects required to meet lead and zinc demand. These supply functions are then incorporated into econometric industry models that include details on lead demand, inventories, secondary supply, scrap, and prices for the major countries and geographic regions. The model was employed to generate medium to long-term regional projections based on expected market prospects, supply outlooks, costs of production, prices, and forward-linked industrial demands. Because mineral market fluctuations occur mostly on the demand side, Newcomb, et.al. (1990) have shown how stochastically varying demands can be introduced in their model of world aluminum trade.

One of the first applications of linear programming to analyze spatial allocation problems in energy markets was made by Henderson (1958) who analyzed competitiveness in the coal industry. Coal demands and supplies were identified among fourteen regions in the United States. The objective function which was minimized featured the delivered costs (i.e., extraction plus transportation costs) of coal allocation patterns, subject to competitive market conditions. Deviations from competitive efficiency were then evaluated by comparing the efficient model solution with existing coal allocation patterns.

When coal again gained stature as a prominent energy source in the 1970s, a more elaborate linear programming model comparing the supply and demand potential of Western coals compared to Eastern coals in the United States was constructed by Libbin and Boehji (1977). Their model expanded regional coal supply activities to include methods of surface and underground mining. Three coal quality levels based on sulfur content and heating value were incorporated, and their demand allocations also considered the blending of coals to meet emissions sulfur standards. Solution of the model minimized the discounted total cost of meeting national coal demands, subject to variations in sulfur burning standards. The results were interpreted regionally in terms of the amounts of coal that various Western and Eastern states could supply. More recently, Bernkopf (1985) extended this approach to regions throughout the United States. In addition, Quirk, Katsuki, and Whipple (1982) reviewed national coal model applications from a government planning perspective.

One of the more elaborate attempts in the late 1970s to use linear programming in spatial and process energy analysis was the Project Independence Evaluation System (PIES) constructed for the Federal Energy Agency FEA) (now the Energy Information Administration [EIA]). This model (EIA, 1979) includes a macroeconometric model, an econometric demand model, and a linear programming model that explains fuel supplies, conversion, and shipments. The PIES model was started under the Nixon administration to analyze a policy of energy independence. It was later shelved during the Ford administration and subsequently renewed under the Carter administration. Later, it received severe criticism by Commoner (1983).

The solution of the PIES model depends on an integrating model that uses given estimates of regional demands, prices and elasticities, regional supply schedules, and resource input requirements to calculate an energy market equilibrium. The linkages between the demand model and the linear programming submodel, which incorporate the supply schedules and conversion processes, depend on the computation of a price/quantity coordinate on the demand curve for each of the primary and derived energy products in the system. Associated with each of these coordinates are measures of the sensitivity of the quantities demanded to small changes in each of the prices in the demand model (own- and cross-price elasticities). The linear programming problem is solved so that the minimum cost schedule of production, distribution, and transportation necessary to satisfy the given demand levels is reached. This process continues until the demand and supply prices are equal, at which point the energy market is assumed to be in equilibrium. One useful aspect of the PIES effort is the combined or integrating model algorithm developed by Hogan and Weyant (1980). While benefiting other similar modeling efforts, such a simultaneous model solution can be easily obtained without integration by alternatively using spatial and temporal price and allocation models, e.g., see Takayama (1979).

There are other applications of linear programming to energy modeling. Deam et al. (1974) examined patterns of petroleum trade by building a spatial energy model in which the petroleum/natural gas component dominates, and Devanney and Kennedy (1980) paid particular attention to world refinery locations and flows. Miranda and Glauber (1988) provide insights into adapting the spatial equilibrium model for the possible influence of trade on production uncertainty in the petroleum industry. Most of these models combine regional, process, and demand characteristics, making model hybridization one of the principle outcomes of this era of linear programming applications.

Other recent applications have examined specific regulatory, policy-oriented issues in relation to environmental controls. For example, Schlottman and Watson (1989) analyze coal shipments between U.S. supply and demand regions according to air quality standards and consequent reduction of sulfur emissions from coal conversion in electric power plants. Suwala et al. (1997) have developed a similar modeling approach to explain the impacts of emission standards on the Polish coal industry. Similarly,

Provenzano (1989) analyzes changes in regional energy facility location patterns that might be induced by changes in national energy development policies or changes in regional water allocation requirements. Regional water resource development is thus linked to regional energy development when examining substitution between nuclear and coal electricity generation.

PROCESS PROGRAMMING

Process programming models recognize an important characteristic of mineral trade: that the commodities used and traded within industrial requirements normally involve different stages of process or production e.g., see Kovisars (1976). These stages can include mining, ore treatment (milling and concentration), reduction (smelting), purification (refining), and consumption by fabricators. Recycled material also is an important process input for many mineral flows and may enter the supply flow at several stages. Manne and Markowitz (1963) described the early development of these models as an application of process analysis, but they also suggested possibilities for integrating this approach with spatial equilibrium analysis. Fox (1963) presented such a process model of spatial equilibrium for agriculture, and Marschak (1963) developed one for petroleum refining. Further development of this method is described in surveys by Sparrow and Soyster (1980) and the National Academy of Sciences (1982); both recognize process programming as a distinct class of mineral and energy models.

The actual formulation of a process programming model is normally based on linear programming, although other forms of mathematical programming can be used. Solution of the model usually involves minimizing an objective function specified in terms of production costs, subject to constraints such as the time sequence of production, regional capacities, the demand for products, and technical relationships among the production variables. An activity component distinguishes process programming from other models. That additional component describes how decision variables (resources in the production problem) are combined in fixed proportions by the production technologies to produce an output. This emphasis makes process programming, in some ways, similar to the input-output approach, but it allows multiple production processes or activities. These activities are technologically possible alternatives in physical terms (i.e., different energy and material requirements and labor requirements) and do not necessarily yield economically (or socially) efficient solutions.

Economic considerations are introduced by a cost function, which shows the minimum cost of producing various levels of output, given factor prices and technologies.

Process models are well suited for conducting mineral supply analysis in a regional context. Usually formulated in terms of production technologies, they include labor, energy, and materials requirement inputs at each stage of the production sequence. Mathematical equations represent the various technological or engineering production possibilities. The overall production flow is disaggregated into elementary process routes, and input-output parameters based on engineering data can be derived for each stage of each route. For each process, programming techniques select the production possibilities that optimize a particular goal.

A mineral process model permits demand analysis by linking final product demands and derived material demands. National economic activity can be used to drive final product demands, with the process model then explaining how primary products derived from materials can be transformed into secondary, tertiary, and other products until final product demand is met. Sometimes the supply of a material is also studied as part of the demand process.

Ray and Szekely (1973) searched for the least-cost combination of production technologies in an integrated steel plant model, subject to demand, technology, and raw material constraints. Their formulation finds the optimal mix of three process routes (blast furnace, open hearth furnace, or electric furnace) depending on the prices of the factor inputs, such as scrap and hot metal. Such a formulation is useful for studying the demand for materials that are used in the production sequence.

Another process programming model of the steel industry was built by Clark and Church (1981). Their model represents the production of stainless steel products by alternative technological routes starting with basic raw materials, such as carbon and stainless steel scrap, ferrochromium, ferronickel, and other materials. This type of model can be used to analyze the demand for input materials required in the production process. For instance, in the case of stainless steel, the demand for chromium, nickel, manganese, and other materials can be estimated as a function of the production of stainless steel by different processing routes.

In addition to Kovisars' applications (1975, 1976), which extended process analysis to spatial mineral modeling, such models have also dealt with the international petroleum refining industry. Devanney and Kennedy (1980) created a hybridized energy model as briefly mentioned in the discussion of linear programming. Their process activities include the production of crude oil, its transformation to refinery products, and the consumption of these products. The model has seven regions (the United States, Canada, Latin America, Western Europe, the Middle East, Africa, and Asia) and determines the optimal spatial allocation of the products, subject to government regulations, such as tariffs, excise taxes, environmental restrictions, refinery subsidies, and policies intended to shift the supply or demand curves.

A process model of the world oil market by Deam (1974) included fuel transportation among 25 worldwide geographical regions. Fifty-two types of crude oil and twenty-two refining centers are represented along with six types of tankers. The linear programming matrix for this model has about 2,500 rows and 13,500 columns. The exogenous inputs to the model include the regional demand for products, refinery technologies, costs of product refining, and transport of specific crudes and products. The model determines the optimal allocation of the routing of crude oil and products between sources, refineries, tankers, and production facilities to satisfy the projected levels and distribution of demands. Because the model includes the transport and refining costs for crude from specific sources, it can analyze the relative price of these crudes in a competitive market or in a market where relative prices are set to reflect the differences in transportation and refining costs among the many sources.

QUADRATIC PROGRAMMING

Quadratic spatial models determine demand, supply and prices endogenously in a simultaneous context. As developed by Takayama and Judge (1971), the quadratic programming formulation features (1) a system of equations describing the aggregate demand and supply for one or more commodities in different demand and supply regions, (2) the distribution activities over these regions, and (3) the market equilibrium conditions. Although the demand and supply equations imply a structure similar to that of a temporal market model, the equilibrium process identifies profits to be realized from the flow of commodities, i.e., the price differential between two regional points minus transportation costs. Profit maximization is assured by a computational algorithm that transfers commodities until demand equals supply in every spatially separated market. To evaluate policy decisions, the equilibrium conditions and other definitional equations can be used to impose constraints on the model parameters.

The structure of an elementary spatial and temporal price and allocation model (STPA) is described in the following example, which uses a quadratic objective function. Linearity is assumed in demand and supply and the necessary identifying variables are embodied in the constant terms.

$$D_i = b_{0i} + b_{1i} P_i \quad \text{for all } i \tag{5.9}$$

$$S_j = b_{0j} + b_{1j} P_j \quad \text{for all } j \tag{5.10}$$

where $b_{1i} < 0$ and $b_{1j} > 0$; P_i is the commodity demand price in region i; and P_j is the commodity supply price in region j. Takayama and Judge (1971) express these equations in their inverse form, so that equations (5.9) and (5.10) can be rewritten as

$$P_i = a_{1i} + a_{2i} D_i \quad \text{for all i}$$
$$P_j = a_{3j} + a_{4j} S_j \quad \text{for all j}$$

where a_{1i}, a_{3j}, $a_{4j} > 0$ and $a_{2j} < 0$ over all observations. The constraints imposed on supply and demand are the same as in the linear programming model.

$$D_i \leq \sum_j^n Q_{ij} \quad \text{for all i} \tag{5.11}$$

$$S_j \geq \sum_j^n Q_{ij} \quad \text{for all j} \tag{5.12}$$

Transport costs and shipments are assumed to be non-negative,

$$T_{ij}, Q_{ij} \geq 0. \tag{5.13}$$

The objective function necessary to complete the model goes beyond the cost minimization goal of linear programming. It maximizes the global sum of producers' and consumers' surplus after the deduction of transportation costs. This form of market-oriented quasi-welfare function has been termed net social payoff (NSP) by Samuelson (1952) and is defined by:

$$NSP = \sum_{i}^{n} \int_{0}^{D} P_i(D_i)dD_i - \sum_{j}^{n} \int_{0}^{S} P_j(S_j)dS_j$$
$$- \sum_{i}^{n} \sum_{j}^{n} Q_{ij}T_{ij} \tag{5.14}$$

The objective function in this example (5.14) can be rewritten after substituting the linear demand and supply relations as follows

$$Max\,(NSP) = \sum_{i}^{n} a_{1i} D_i - \sum_{j}^{n} a_{3j} S_j$$
$$- 1/2 \sum_{i}^{n} a_{2i} D_i^2 - 1/2 \sum_{j}^{n} a_{4j} S_j^2 - \sum_{i}^{n} \sum_{j}^{n} T_{ij} Q_{ij} \tag{5.15}$$

Temporal behavior can be introduced by solving the model over time.

Time-varying parameters can be introduced, and separable or other programming algorithms can be employed. As the models' size increases, larger nonlinear programming models can be broken into submodels. The number of constraints, such as the commodity flow restrictions in the model, can be increased to make the number of inter-regional trade flows realistic. The basic solution of the model specifies that the number of flows cannot exceed one less than the number of consuming plus producing regions in the model. More recent interpretations of space-time patterns for the price variable appear in Roehner (1995).

While mineral model applications of quadratic programming have been limited, energy model applications have appeared more frequently. For example, Uri (1976) has examined the efficiency with which electricity is generated and allocated in the United States. His model characterized the electrical industry as having separate supply and demand locations and a fixed amount of transmission capacity at any one time. He selected a partial equilibrium solution because other energy markets, although exogenous to the electricity market, affect overall energy market equilibrium. His formulation assumes that a unified authority responsible for the allocation and pricing of electrical energy acts within a competitive market to maximize social welfare in relation to available and future electrical energy supplies. One of his innovations was to introduce monopoly considerations by establishing a price structure that specifies differing elasticities of demand among various categories of consumers. This form of price discrimination permits the utility to recover full-costs plus its allowable rate of return. Social welfare losses (decreases in net social payoff) are thus examined in relation to this consumer pricing system. Uri (1976, 1977) later showed how industry capacity could be varied within the model solution by introducing concepts of investment and capital formation.

Quadratic programming applications also have been made to the domestic and the international coal market. Labys and Yang (1980) and Yang and Labys (1981) used this approach to model the regional allocation of Appalachian coal shipments. Their model begins with a regional allocation limited to Appalachian steam coal supplying eastern steam coal demand regions. It improved on the supply and demand allocations of the linear programming models by including econometric demand and supply equations so that coal quantities and prices are solved simultaneously in the model. With an objective function maximizing net social payoff, the sensitivity of regional coal demand and supply is measured in response to changes in price elasticities, transportation costs, and ad-valorem tax rates. Yang and Labys (1985) later extended this model to enable substitution between coal and natural gas. Newcomb and Fan (1980) supplemented the model by adding geological factors in the supply equations. A slightly different model was also developed by Campbell et al. (1980) which included all U.S. coal supply regions. At the international level, Dutton (1982) applied quadratic programming to international coal trade in order to assess the impact of future coal prices on regional import policies of selected coal consuming countries.

Kennedy (1974) also applied quadratic programming at the world level. His model was built to prepare simulation analyses and forecasts of regional flows in the international crude oil and refinery market. He examined the impacts of government policies and changing exogenous factors, such as tanker technology and the cost of finding and producing oil in more remote regions on regional trade, environmental conditions, and taxation. For each region and fuel or refined product, the model determines the levels of production and consumption, prices, product refinery capital structure, and world oil trade flows. Takayama (1979) also has confirmed the usefulness of the quadratic programming approach for assessing world energy markets.

It should be noted that quadratic programming avoids the cumbersome integrating routines between demand and supply components that were employed in the above PIES model (EIA, 1979). This advantage has been demonstrated in Hashimoto's (1977) model, which combines separate world food and energy models and generates regional solutions on the basis of related macroeconomic model forecasts.

MIXED INTEGER PROGRAMMING

The multi-period linear mixed integer programming model (MIP) of the type developed by Kendrick (1967) and Kendrick and Stoutjesdijk (1978) combines attributes of both the spatial and intertemporal equilibrium models. Like spatial equilibrium, it explains the flows of commodities by stage of process between regions subject to price differences and costs. Like recursive programming (Day, 1973), it is a dynamic application of linear programming, except that an integer characteristic is introduced to accommodate combinations of binary variables, e.g., the nonexistence or existence of a production facility. Mixed integer programming stems from attempts to cope with several commodity-oriented analyses at once, including regional shipping and transportation, industrial process, intertemporal considerations, or investment project selection. Several of these functions are typically combined to analyze market adjustments, as exemplified in Dammert's (1980) study of copper investment in Latin America. Programming can also be facilitated using the GAMS software by Brooke et al. (1988). This package also features a number of mineral and energy models of this type as well as nonlinear formulations.

Specifying a mixed programming model begins with the regional transport component, which resembles the spatial equilibrium transportation model discussed earlier. A process component is added to deal with the variety of mineral or energy products. And a project selection component incorporates investment to augment capacity, to gain economies of scale, or to increase exports. Kendrick and Stoutjesdijk (1980, pp. 50-55) provide an example of a mixed integer programming model applied to mineral industry development. They attempt to find the minimum discounted cost of meeting specified market requirements over a given period. This search involves the specification of activity levels for increments to capacity; these include shipments from plants to markets and among plants; imports and exports; and domestic purchases of raw materials, miscellaneous material inputs, and labor. Because of the relative complexity of their model, it is characterized here only according to its objective function, cost components, and model constraints.

Their objective function minimizes the total discounted costs of production TC, which consist of capital costs CK, recurrent costs CR, and transport costs CT.

$$Minimize \ TC_t = \sum_t^T \delta_t (CK_t + CR_t + CT_t) \qquad (5.16)$$

The capital cost component equals fixed charges plus the linear portion of capital costs

$$CK_t = \sum_{\lambda}^{t} \sum_{i}^{I} \sum_{m}^{M} \sigma_m (\omega_{mi\lambda} Y_{mi\lambda} + v_{mi\lambda} H_{mi\lambda}) \qquad (5.17)$$

Here λ is a time interval, i is a plant site and I is total sites, m is a productive unit and M is total units, σ_m is the capital recovery factor for m, ω is the fixed charge portion of investment costs, Y represents binary investment decisions, v is the linear portion of investment costs, and H represents continuous investment decisions. The recurrent costs consist of costs related to capacity plus local raw materials and labor costs:

$$CR_t = \sum_{\lambda=1}^{t} \sum_{i}^{I} \sum_{m}^{M} \beta_{mi\lambda} \overline{H}_{mi\lambda} + \sum_{C}^{CR} \sum_{i}^{I} \pi_{cit} R_{cit} \qquad (5.18)$$

The subscript c defines commodities used in the industry and the index CR represents raw materials, miscellaneous inputs, and labor; β is the portion of recurrent costs that is proportional to the capacity; $\overline{H}$ represents maximum capacity expansion per time period; π is price; and R represents domestic purchases of materials and labor.

The regional transport cost component equals final and intermediate product shipment costs. Raw materials prices are assumed to include domestic transport costs:

$$CT_t = \sum_{c}^{CF} \sum_{i}^{I} \sum_{j}^{J} \mu_{cijt} X_{cijt} + \sum_{c}^{CF} \sum_{i}^{I} \mu_{c'iit} X_{c'iit} \quad i' \neq i \qquad (5.19)$$

CT represents final products of the industry, j is domestic market areas, J is total markets, μ is unit transportation costs, and X represents domestic shipments.

At the next step, the production level Q of commodity c by all processes p at plant i must at least equal the shipments X of commodity c from plant i to all markets j. The typical production process that provides final commodities can be assigned a coefficient $a_{cpi} = 1$ in the final commodities constraint, because the unit of capacity can be arbitrarily defined in terms of one of the process inputs (-) or outputs (+)

$$\sum_{p}^{P} \alpha_{cpi} Q_{pit} \geq \sum_{j}^{J} X_{cijt} \qquad (5.20)$$

The output of intermediate commodities at plant i must be greater than or equal to shipments of intermediate commodities from plant i to plant j

$$\sum_{\pi}^{\Pi} \alpha_{\chi\pi\iota} \Theta_{\pi\iota\tau} \geq \sum_{\iota}^{I} \Xi_{\chi\iota\iota,\tau} \qquad \iota \neq \iota' \qquad (5.21)$$

The production of intermediate and final products requires raw materials and labor. The coefficient α_{cpi} in constraint (5.20), thus, will normally be negative. Purchases of raw materials and labor, R_{ci}, will then have to be positive for the constraint to hold

$$\sum_{p}^{P} \alpha_{cpi} Q_{\pi} + R_{ci} \geq 0 \qquad (5.22)$$

Capacity required is less than or equal to expansion less capacity retirements

$$\sum_{p}^{P} b_{mpi} Q_{pit} \leq K_{mi} + \sum_{\lambda}^{T} (\overline{H}_{mi\lambda} - S_{mi\lambda}) \qquad \lambda \leq t \quad (5.23)$$

Here, b_{mpi} is the unit of capacity used on productive unit m per unit of output of process p; K_{mi} is initial capacity for productive unit m at plant I, and S_{mi} is the expected retirement of capacity for productive unit m at plant I in time period λ. The S variables are closed exogenously to the model. For example, if a productive unit manufactures steel, the initial capacity might include a number of open hearth furnaces that were slated for retirement during the period covered by the model. Then S_{mi} would represent the capacity to be retired in each time period λ. The summation over λ in equation (5.23) for λ less than or equal to t permits all capacity installed in previous periods to be available for use in period t.

Two constraints are used to complete the specification of investment in the model.

$$H_{mit} \geq \overline{H}_{mit} \, Y_{mit}, \quad and \qquad (5.24)$$

$$Y_{mit} = 0 \ or \ 1$$

where $\overline{H}_{mit}$ is an upper bound on the size of capacity unit that can be added to productive unit m at plant I in period t. These constraints introduce the integer conditions directly into the model:

$$Y = 0 \text{ when } H = 0, \text{ and}$$

$$Y = 1 \text{ when } H > 0.$$

The effect of equations (5.24) is to prohibit any addition to capacity unless a fixed charge is incurred; however, a fixed charge is only incurred if Y_{mit} is equal to 1. From equations (5.24) for the constraint to hold, Y_{mit} must equal 1 if H_{mit} is positive. If H_{mit} is 0, the model forces Y_{mit} to 0 because the cost minimization objective leads to a preference not to incur the fixed charge.

The summation of regional shipments from all plants I to each market j must be equal to or greater than the product requirement of market j:

$$\sum_{i}^{I} X_{cij} \geq W_{cj} \qquad (5.25)$$

Finally, there are the non-negativity constraints:

$$X_{cij}, Q_{pi}, W_{cj}, H_{mi}, R_{ci}, S_{mi} \geq 0$$

Applications of the mixed integer programming model have taken advantage of the integer constraint to model investment in mineral and energy industries. Important issues such as the determination of efficient regional investment patterns, project and program evaluation, regional economic integration plans, and industry regulations can be explored. Choksi, Meerhaus, and Stoutjesdijk (1983) provide an example of a comprehensive application to regional investment and production allocation in the world fertilizer industry. Their principal goal was to determine the optimal locations of fertilizer plants in different countries, given mineral and gas feedstock availabilities. Variables optimized include production (by-products), shipments, and domestic demands and exports.

Dammert and Palaniappan (1985) used MIP to model the world copper market and copper investment planning in Latin America. Their approach included reserve levels and reserve limits together with mineral exhaustion through a constraint on processing different ore grades:

$$Z_{pit} \leq q_{cpit} \tag{5.26}$$

This constraint places an upper limit on Z, the annual exploitation of high-grade and second-grade ores, where q_{cpit} is the annual availability of ore grade c. It can be extended to include the total reserve limit

$$\sum_{t}^{T} Z_{pit} \leq v_{pi} \tag{5.27}$$

where v_{pi} is the total reserves of ore grade in processing area i.

The study by Brown et al. (1983), which also involved Dammert, Meeraus, and Stoutjesdijk, provides an investment analysis of the aluminum industry similar to the one for copper. The objective was to minimize overall investment and operating and transportation costs to meet market requirements. Investment decisions are expressed in an integer (0-1) format, and the depletion of bauxite mine reserves is also considered. Energy costs play an important role in determining investment location to meet the then long run aluminum demand in the year 2000.

Kendrick, Meeraus, and Alatorre (1984) constructed an MIP model of the steel industry, and Kendrick's students at the University of Texas have also employed MIP to a model variety of energy problems. Examples include the Gulf Coast refining complex (Langston, 1983), the Korean electric power industry (Kwang, 1981), and the Korean petrochemical industry (Jung, 1982).

LINEAR COMPLEMENTARITY PROGRAMMING

The standard quadratic programming approach requires symmetry of its regression coefficients, a condition sometimes referred to as the integrability condition. However, there is no reason why independently estimated final goods demand or supply functions for each region have to satisfy this condition. The likelihood of such a symmetrical relation between each pair of commodities is remote, if not impossible. Linear restrictions could be imposed on the coefficients to preserve symmetry, but an artificial spatial equilibrium would result. To overcome this situation, the linear complementarity programming approach of Cottle and Dantzig (1968) was developed to solve commodity spatial problems. It frees the spatial equilibrium model from being normative and restrictive and maintains the efficiency of its quadratic programming counterpart. Although the linear complimentarity programming model does not maximize net social payoff, that concept is not always important for model applications. Instead, this model optimizes revenues or costs consistent with a set of operational rules that follow the underlying Kuhn-Tucker (1951) conditions.

The theoretical linear complementarity programming structures conforming to this formulation are reviewed in Takayama and Labys (1986) and Takayama and Hashimoto (1984). The following problem formulation based on Yang and Labys (1985) deals with the simple example of the regional demand and supply allocation of two commodities. Equations for the demand and supply of these commodities can be expressed in linear form.

$$Pd_j^1 = \alpha_j^1 + b_j^{11} y_j^1 + b_j^{12} y_j^2$$

$$Pd_j^2 = \alpha_j^2 + b_j^{22} y_j^2 + b_j^{21} y_j^1$$

$$Ps_i^1 = e_i^1 + f_i^{11} x_i^1 + f_1^{12} x_i^2$$

$$Ps_i^2 = e_i^2 + f_i^{22} x_i^2 + f_i^{21} x_i^1$$

(5.28)

Here the superscripts 1 and 2 refer to the two separate commodities; the subscripts denote supply region i and demand region j; y and x denote consumption and production, respectively; and P_d and P_s denote the demand price and the supply price, respectively. For instance, P_s^1 (supply price of commodity 1 in region i) is a linear function of X_i^1 (quantity supplied of commodity 1 in region i) and of X_i^2 (quantity supplied of commodity 2 in region i).

The basic model seeks a nonnegative solution for the vector formulation given by X, Y, Z, λ, and γ, subject to the following equilibrium condition.

$$\gamma \leq Ps\,(X) \quad \text{where} \quad X^T\,[Ps(X) - \gamma] = 0 \qquad (5.29)$$

$$\lambda \geq Pd\,(Y) \quad \text{where} \quad Y^T\,[\lambda - Pd(Y)] = 0 \qquad (5.30)$$

$$G^T\begin{bmatrix}\lambda \\ \gamma\end{bmatrix} - T \leq 0 \qquad \left(G^T\begin{bmatrix}\lambda \\ \gamma\end{bmatrix} - T\right)^T Z = 0 \qquad (5.31)$$

$$GZ - \begin{bmatrix}Y \\ -X\end{bmatrix} \geq 0 \qquad \left(GZ - \begin{bmatrix}Y \\ -X\end{bmatrix}\right)^T \begin{bmatrix}\lambda \\ \gamma\end{bmatrix} = 0 \qquad (5.32)$$

Here X and γ are elements of R^{mk}, Y and λ are elements of R^{mk}, Z and T are elements of $R^{mn,k}$, and G is an element of $R^{k(m+n),kmn}$. Ps(X) and Pd(Y) are column vectors in $R^{m,k}$ and $R^{n,k}$, respectively, and denote observed supply and demand prices; m and n denote the number of supply and demand regions; and k denotes the number of commodities. Superscript T is the conventional transpose, and G is a flow coefficient matrix.

One way to explain linear complementarity programming is in terms of its relation to quadratic programming and linear programming. The model formulation would require that vectors w and z satisfy the following conditions

$$w = q + Mz$$
$$z^T w = 0 \qquad (5.33)$$
$$w \geq 0$$
$$and \ \ z \geq 0$$

where **q** is a real (p x 1) vector, **M** is a real (p x p) matrix, and **w** and **z** are (p x 1) vectors. Bimatrix (two-person nonzero-sum) games were defined in this form (Lemke, 1965), but the problem is defined as f programming because linear complementarity programming, as defined later, is a special case of programming.

The quadratic programming model equivalently defined requires an (n x 1) vector x that maximizes

$$Max\ L = c^T x - 1/2\ x^T Qx \qquad (5.34)$$

subject to

$$Ax\ \leq b,\ \ and$$

$$x \geq 0$$

where c is an (n x 1) real vector, Q an (n x n) real matrix, A an (m x n) real matrix, and b an (m x 1) real vector. This quadratic programming problem can be converted to the form of Kuhn and Tucker (1950), which makes it more directly equivalent.

If we consider the more general formulation of nonlinear complementarity programming, then the following inclusion relationships hold: Nonlinear complementarity programming contains or is equal to linear complementarity programming, which contains or is equal to quadratic programming. Linear programming can then be considered a special case of quadratic programming where Q = 0 and is a subset of quadratic programming. Hence, the following inclusion relationships are established: nonlinear complementarity programming contains linear complementarity programming, which contains quadratic programming, which contains linear programming. Linear complementarity programming, as a special case of nonlinear complementarity programming, can deal with a wider range of issues than linear programming. The most elementary linear complementarity programming application takes advantage of the asymmetry in the parameters of the demand equations for the commodity or commodities of interest. Yang and Labys (1985) have examined the potential for employing linear complementarity programming solutions to modify their spatial and temporal price and allocation model of the Appalachian coal industry. The new model now includes natural gas and coal in the spatial allocation of the Appalachian coal and gas market, and searches for the optimal trade flows of gas and coal among the major Appalachian supply and eastern demand regions. Although the model is static, providing only a one-period solution, full simultaneity is obtained in the determination of equilibrium quantities and prices.

More dramatic applications of linear complementarity programming provide intertemporal linkages through investment determination, for instance, as in the Hashimoto and Sihsobhon (1981) iron and steel model. The major advance of their model is its incorporation of market expectations based on forward information and market dynamics. One version assumes that the steel industry plans and implements investments in production facilities rationally, with perfect foresight. Another version assumes that the industry follows less than perfect investment plans. Model solutions give projections for the following variables: (1) investments in steel production capacities, (2) prices, demand and supply quantities for steel products, and steel production capacities, and (3) the industry's requirements for major raw materials and inputs. The results help to explain the kinds of output and investment cycles that have characterized that industry.

Going beyond linear complementarily programming, Nagurney (1987) has shown how the variational inequality approach can be combined with network theory to study spatial price equilibrium problems, specifically policy interventions considered as disequilibrium, or constrained equilibrium problems. Although the perfectly competitive spatial price problem is addressed within an equilibrium/disequilibrium framework, the variational inequality problem has been specified to contain not only those problems but also minimization problems and virtually all of the classical problems of mathematical programming, such as linear and convex programming, and linear complementarity problems. In addition, it constitutes an alternative approach to fixed point problems, minimax problems, and noncomplementarity problems, i.e. also see Labys and Yang (1996).

6 RESOURCE EXHAUSTION MODELS

Resource exhaustion models represent a special class of mineral and energy models that constitute extensions of econometric models and programming models as well. Since programming models inherently feature optimization, the latter are considered additionally here as an extension of the controlled-market models defined earlier. To better explain this methodology, we begin with the previously given monopolist models and then advance to Stackelberg, Nash-Cournot and modified optimization models. While resource exhaustion models can be constructed for competitive markets, the analysis is restricted primarily to the noncompetitive case of crude oil. Nonetheless, other market and commodity configurations are possible.

MONOPOLIST MODELS

Optimization models of resource exhaustion that describe the crude oil market as consisting of a monopolist or cartel and a competitive fringe resemble the controlled econometric market models described earlier. The approach of Pindyck (1978a,b) describes net demand facing the cartel as

$$X_t = D_t - S_t \qquad (6.1)$$

where D is total market demand and S is the supply of the competitive fringe

$$D_t = d(D_{t-1}, P_t, A_t) \qquad (6.2)$$

$$S_t = s(S_{t-1}, P_t, N_t) \qquad (6.3)$$

exhaustion comes into play for the competitive fringe as well as for the cartel

$$S_t = s(S_{t-1}, P_t, CS_t)$$ (6.4)

where cumulative production CS is given by

$$CS_t = CS_{t-1} + S_t$$ (6.5)

A similar accounting identity is needed to keep track of cartel reserves R.

$$R_t = R_{t-1} - D_t$$ (6.6)

The objective of the cartel is to pick a price trajectory P_t that will maximize the sum of discounted profits

$$\text{Max } W = \sum_{t-1}^{N} (1/ (1+\delta)^t) P_t - m/ R_t \ D_t$$ (6.7)

where m/R is average (and marginal) production costs (so that the parameter m determines initial average costs), δ is the discount rate, and N is chosen to be large enough to approximate the infinite-horizon problem. Note that average costs become infinite as the oil reserve base R approaches zero, so that the resource exhaustion constraint need not be introduced explicitly. The resulting model framework is that of a classical, unconstrained discrete-time optimal control problem, where numerical solutions can be easily obtained.

The control solution to this model yields an optimal price trajectory P^* as well as the optimal sum of discounted profits W^* for the monopolist. One might like to compare these variables with the optimal price trajectory and sum of discounted profits that would result if the cartel dissolved (or never formed), and its member producers behaved competitively. Optimal here implies that competitive producers must manage the exhaustion of their oil reserves over time, balancing profits this year against profits in future years.

Although competitive producers cannot collectively set price, they each determine output, at a given price. Pindyck shows that the rate of output should be such that the competitive price satisfies the equation

$$P_t = (1 + \delta) P_{t-1} - m/R_{t-1} \qquad (6.8)$$

If this were not the case, larger profits could be obtained by shifting output from one period to another. In addition, the initial price must be such that two constraints hold. First, the resulting price P and output D trajectories must both satisfy net demand at every point in time as given by equations (6.1), (6.2), and (6.3), i.e., supply and demand must be in market equilibrium. Second, as the price rises monotonically over time, the exhaustion of oil reserves must occur at the same time that net demand goes to zero. If demand becomes zero before exhaustion occurs, some of the oil would be wasted and would yield zero profits; profits would be greater if the oil were depleted more rapidly (at a lower price). If exhaustion occurs before demand becomes zero, depletion is occurring too rapidly and should proceed more slowly.

The computation of the optimal price trajectory for the competitive case is thus straightforward. Pick an initial P_o and solve equation (6.7) over time together with equations (6.1), (6.2), (6.3), and (6.6). Repeat this for different values of P_o until D_t and R_t become zero simultaneously. Results of application of such optimizing models to OPEC behavior can be found in Hnyilicza and Pindyck (1976) as well as in Cremer and Weitzman (1976).

STACKELBERG MODELS

A first alternative to the monopolist model is to assume some interaction between the monopolist and the fringe, i.e., the case of a nonuniform cartel. This changes the model theory from that of monopoly to duopoly. The theoretical approach for explaining oil market and price behavior in this context is the Stackelberg model of the dominant firm in which the latter takes the reaction of other firms into account in its pricing policy, while the fringe or other firms accept prices as given. Such an oil model reflects the theory of dynamic limit pricing. The cartel chooses a production path or pricing policy that maximizes net revenues, and those net revenues depend on the rate of production by the competitive fringe. Unlike classical dynamic limit-pricing models, where the residual demand of the dominant firm depends only on its current rate of production, the response of the competitive fringe is a function of the entire sequence of outputs determined by the oil cartel.

Let us examine the simple example of a Stackelberg oil model presented in Aperjis (1981). OPEC is divided into two groups, one of which can be said to be dominant over the other. Using the distinction of foreign exchange absorption, the low absorber group has the largest oil reserves in OPEC and consequently, the largest potential to expand its productive capacity. The higher absorber group has less reserves and views the cost of an OPEC breakup as being much higher to them than to the low absorbers, because the former group has a relatively greater need for oil revenues.

Define Group 1 as the low absorbers and Group 2 as the high absorbers and assume that Group 1 dominates Group 2. According to the Stackelberg model, Group 1 will be an oil price-setter and quantity-follower, while Group 2 will be an oil price-taker and quantity-setter. In other words, Group 1 sets oil prices in such a way as to maximize its profits, while Group 2 takes these prices as given and sets oil production at a level which maximizes its profits. Finally, Group 1 produces enough to satisfy any residual demand for OPEC oil over that met by the production of Group 2.

A model of this behavior can be formulated in the following way. Let Group 1 determine the sequence of oil prices P^1 which maximizes its profits according to the following maximization problem

$$Maximize \sum_{t=1}^{N\,I} [P_t^1 - C_t^1] \, X_t (1/(1+\delta_1)^t)] \tag{6.9}$$

subject to

$$\sum_{t=1}^{NI} X_t \leq R_t^1 \tag{6.10}$$

with the solution

$$MR_t^1 = MC_t^1 + \lambda^1 (1+\delta_1)^t \tag{6.11}$$

Subsequently, Group 2 takes oil prices P^2 and produces at that rate S which maximizes its profits according to

$$Maximize \sum_{t=1}^{N} 2 \, [P_t^2 - C_t^2] \, S_t [1/(1+\delta_2)^t] \tag{6.12}$$

subject to

$$\sum_{t=1}^{N2} S_t \leq R^2 t \qquad (6.13)$$

with the solution

$$P_t^2 = MC_t^2 + \lambda^2 (1 + \delta_2)^t \qquad (6.14)$$

where:

P^1, P^2	= Prices of Groups 1, 2
C^1, C^2	= Unit production costs of Groups 1, 2
δ_1, δ_2	= Discounts of Groups 1, 2
X	= Production of Group 1
S	= Production of Group 2
R^1, R^2	= Reserves of Groups 1, 2
MR^1, MR^2	= Marginal revenue of Groups 1, 2
MC^1, MC^2	= Marginal costs of Groups 1, 2
λ^1, λ^2	= Foregone future opportunity of an additional unit of current production of Groups 1, 2

The recognition of Group 2 by Group 1 (the cartel) will cause the latter to choose a price trajectory different from the previous monopoly model, both in magnitude and in rate of change over time. The implications of this theoretical approach can be seen, for example, in the oil modeling study of Gilbert (1978). In addition, MacFadden has investigated the impact of changing the nature of the oligopolistic kinked demand curve to analyze OPEC's behavior.

NASH-COURNOT MODELS

A second alternative to the oil monopolist model is to assume duopoly market behavior in the form of a non-uniform cartel but to change the behavioral pattern from that of followers to that of bargaining or gaming. Referring to the previous example, if the low absorber group is not willing to be a quantity-follower and the higher absorber not willing to a price-taker, then a conflict arises which can only be resolved through a process of bargaining.

Nash (1953) developed a solution to such cooperative games. As shown by Hnyilicza and Pindyck (1976), the two-part cartel can again be described in terms of the behavior of Group 1 and Group 2 with the following objectives.

$$Maximize \ W_1 = \sum_{t=1}^{N1} [P_t - m_1 / R_t^1] X_t^1 (1/(1+\delta_1)^t) \qquad (6.15)$$

$$Maximize \ W_2 = \sum_{t=1}^{n1} [P_t - m_2 / R_t^2] X_t^2 (1/(1+\delta_2)^t) \qquad (6.16)$$

Here δ_1 is assumed to be smaller than δ_2, and X_t^1 and X_t^2 are the production levels of each group. The latter are determined by a division of total cartel production according to

$$X_t^1 = \beta_t X_t \qquad (6.17)$$

$$X_t^2 = (1 - \beta_t) X_t \qquad (6.18)$$

with $0 \le \beta_\tau \le 1$. The exhaustion of oil reserve levels for each group is accounted for by the equations

$$R_t^1 = R_{t-1}^1 - X_t^1 \qquad (6.19)$$

$$R_t^2 = R_{t-1}^2 - X_t^2 \qquad (6.20)$$

To these must be added the following:

$$D_t = d(D_{t-1}, P_t, A_t) \qquad (6.21)$$

$$S_t = s(S_{t-1}, P_t, N_t) \qquad (6.22)$$

$$CS_t = CS_{t-1} - S_t \qquad (6.23)$$

$$X_t = D_t - S_t \qquad (6.24)$$

The next step is to determine how the two groups of countries can cooperate to set oil prices, and to divide output in an optimal manner. Suppose a cooperative agreement is worked out whereby the oil price and output shares are set to maximize a weighted sum of the objectives of each group.

$$Maximize\ W = \alpha W_1 + (1 - \alpha)W_2, \quad 0 \leq \beta_t \leq 1, 0 \leq \alpha \leq 1 \qquad (6.25)$$

The solution to this maximization problem offered by Nash (1953) differs from the two previous ones in that the steps to optimize are more complex. Assume a bargaining game where Group 1 and Group 2 attempt to move along the set of bargaining outcomes in opposite directions; then the problem is to determine a meaningful measure of bargaining power for the two groups. Nash's approach was to introduce the notion of a "threat point" or the outcome that would result if negotiations break down and noncooperative behavior follows.

Of course any solution will depend on the bargaining abilities and power of the two groups. Also a bargaining solution might prevail other than the Nash solution. The mathematical optimization problem involves a control solution to the foregoing set of equations, including W_1 and W_2. One obtains a solution process by repeatedly resolving the optimizations for different values of α. Actual solution values for the case of OPEC can be found in the cited work of Hnyilicza and Pinkyck (1976). Another attempt to model OPEC using the Nash model can be found in the works of Salant et al. (1979), Salant et al. (1981), and Kolstad (1982).

MODIFIED OPTIMIZATION MODELS

Several additional optimization approaches remain which attempt to model OPEC behavior based on different degrees of oligopoly and the continuum of intermediate market structures between pure monopoly and perfect competition as well as on specific policies within OPEC. Among these, the most well-known was that of Kuenne (1980). He viewed resource oligopolies as a community of simultaneously competing and cooperative rivals, each of whom follows strategies that reflect its perception of the industry power structure. Those strategies are aimed at achieving a balance of multiple objectives for each rival.

Any model built to emulate this noncompetitive behavior thus should: (1) contain an explicit formulation of the industry's power structure as perceived by each rival; (2) retain the identities of each individual rival; (3) embrace the set of multiple objectives for each rival; and (4) be flexible enough to adapt to the specific personalities, goals, mores, and institutions of each specific industry analyzed.

To make this approach operational, Kuenne (1980) developed a modified nonlinear programming format called "crippled optimization". That is, each rival maximizes joint industry welfare in a manner that is "hobbled" in two dimensions. First, the firm defines an objective function consisting of its own profits plus the profits of each rival after each has been multiplied by a power structure discount factor, which has been termed a "consonance factor". Second, each firm maximizes this function subject to a set of constraints that contains its subordinate objectives and perceived restraints.

Such a model framework has been seen as having several advantages. First, its flexibility permits the goals of any particular oligopolistic group of firms to be adopted, whether they represent constrained profit maximization, target rates of return, increasing market shares, or whatever. Second, the model is not limited to any single goal. It can incorporate the power structure of the oil industry on a binary one-to-one basis, preserving each firm's perceived relationships to every other. Thirdly, the model can preserve the flavor of realistic oligopoly as a blend of rivalry and cooperation, avoiding the extremes of purely game-theoretical or joint profit maximization approaches. Finally, such a model has been shown to be conceptually and practically operational, permitting modifications of the more well-known algorithms. Although the operational framework of such a model structure is too complex to review here, an example can be found in the Kuenne (1980) description of the GENESYS model dealing with OPEC behavior.

7 INPUT-OUTPUT MODELS

Input-output is a conceptualization of the interdependence of economic production units that has come to be used over the years both as a modeling device and as an element in national economic accounting systems. The input-output (I-O) framework as such, cannot be employed directly to model mineral and energy market behavior. However, it does provide a disaggregated view as to how the demand and supply patterns for different mineral and energy commodities relate to the interindustry structure and the aggregate or macroeconomic variables of a national economy. The construction of input-output models thus requires that some mathematical or econometric model be used in conjunction with an appropriately organized I-O table.

The received input-output methodology has influenced mineral and energy modeling in two ways. First, it has made clear the usefulness of depicting the economic system by means of detailed sectors. For such a purpose each producing sector is defined as a component of the system having a homogeneous output for a given technology. Second, it requires that production must satisfy not only final demand but also intermediate demand needed directly and indirectly to yield final demand. The main contribution of the received input-output structure is that it allows the list of final demands to be transformed into a vector of sectoral outputs.

The basic input-output structure can be defined by dividing the intermediate and final demand activities of an economy into a number of sectors, which are arrayed in matrix form, as shown in Figure 3. The distribution of the sales and purchases of each industry is then estimated for each sector during a one-year period. Final demand can be further disaggregated into the components used in the national income accounts and then mineral and energy consumption forecasts made. Thus, the total final demand for the output of an industry can be considered as the sum of those components:

$$F = C + I + G + N \tag{7.1}$$

where the gross output of an industry is the sum of its sales to other industries and to final demand:

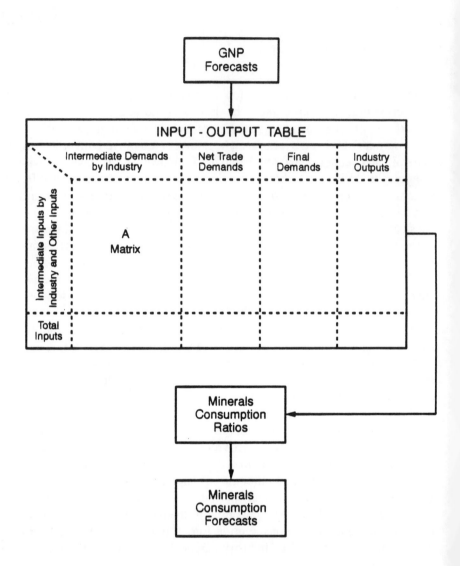

Figure 3

HYPOTHETICAL FORMULATION OF AN INPUT-OUTPUT
MODEL FOR CONSUMPTION FORECASTING

$$Q = SL + F \qquad\qquad (7.2)$$

Definitions:

F	=	Final demand for the output of industry i (i = 1,...,n)
C	=	Personal consumption expenditure for industry i output
I	=	Private investment expenditure for industry i output
G	=	Government expenditure component for industry i output
N	=	Net export (exports minus imports) of industry i output
Q	=	Gross output of industry i
S	=	Sales of industry i to industry j (j = 1,...,n)
L	=	Unit vector of dimension n.

Analogously, the gross output of an industry is the sum of its purchases from other industries and of value added:

$$Q = S'L = V \qquad\qquad (7.3)$$

where V = value added by industry i. Gross national product is measured as the sum of final demand (expenditure approach) or the sum of value added (income approach).

Up to this point, the input-output table has been described as a system of accounting identities. However, in situations where producers are regarded as having only a limited choice regarding factor input intensities and where adjustments to shifts in demand take the form of output adjustments rather than of price adjustments, the transactions table can be utilized to develop a general set of production or technical coefficients from the S matrix. As explained above, a technical coefficient is defined as the dollar input purchases from industry i per dollar output from industry j or

$$A = [\, a_{ij} \,] \qquad\qquad (7.4)$$

where $a_{ij} = S_i / Q_j$

Substituting the value of Q into equation (7.2) yields the results:

$$Q = AQ + F \qquad\qquad (7.5)$$

This is equivalent to

$$(I - A)Q = F \qquad (7.6)$$

where I = an identity matrix. From equation (4.6), one can find the "total requirements matrix,"

$$Q = BF \qquad (7.7)$$

where

$$B = [\, b_{ij} \,] = [I - A\,]^{-1} \qquad (7.8)$$

Each b_{ij} represents the dollar output of industry i required, both directly and indirectly, per unit value of final demand from industry j.

There are several important aspects of applying the completed tables. First of all, use of an I-O table for the economic analysis of mineral and energy commodities is limited largely by the national accounting purposes for which the tables are intended. That is, the more the tables include detailed mineral and energy industries, the better will either or both of these sectors be modeled. Second, an I-O table is organized on the basis of value transactions, including the technical coefficients of the A matrix. While the value (or currency) amounts can be used as surrogates for the underlying physical reality, no system of price deflation exists, for example, that will both preserve the physical constancy of dollars within the input-output cells over time and yet reproduce an independently deflated constant currency GDP. Later, an attempt to formulate a resource-oriented I-O model (SEAS) in quantity terms is discussed. There is also the problem that any given physical quantity of material is in reality marked-up (or down) in value as it proceeds through successive stages of processing and manufacture. This produces anomalies, particularly in the case of circulating industrial scrap. Because it is marked down in value before being returned to earlier stages of processing, the recycled scrap shows up as an input at only a fraction of its true value.

Thirdly, I-O tables as currently constructed make no provision for differences in total input requirements according to the particular kind of final demand. Whether a final product is destined for personal consumption, government, inventory, investment, or export, it is assumed to be exactly the same product and to generate the same direct and indirect unit input requirements. To the extent, therefore, that an I-O table seems to reveal differences in demand for mineral or energy commodities that relate to differences in the kind of final demand, this will be because of differences among the various kinds of demand in their respective product mixes. It follows that the amount of the calculated difference in inputs for the different kinds of final demand will depend heavily on the amount and kind of sector detail that a given table offers.

What makes input-output useful for mineral and energy modeling is the fact that its matrices can be treated as a series of producing sector requirements equations. These requirements (the independent variable) are in each case dependent, according to parameters stated in the matrix, upon exogenous final demand variables as well as upon other intermediate industry variables further up the production chain. In addition some of the producing sectors are specifically mineral production and/or processing sectors, i.e. see Bocoum and Labys (1993). Thus some other configuration or levels may be substituted to determine the implications of the alternative exogenous conditions on the output requirements of the mineral or energy sectors.

The effect of assuming different values for the parameters can also be investigated. However, given the requirement for complete consistency between inputs, outputs, and final demand, only the final demand "bill of goods" may in fact be treated as a fully independent set of variables. Other exogenous alterations, though useful for many practical purposes, may be tainted by the inconsistencies they introduce into the overall matrix, i.e. see Rose (1983). A rebalancing of the matrix may be employed to eliminate formal (accounting) inconsistencies, but will not necessarily preserve the kind of consistencies among input-output parameters that was automatically provided in a historical recording.

The coefficients of such a matrix are linear and static, and imply constant returns to scale. Because of this static character, dynamic phenomena such as capital requirements must be handled in auxiliary fashion. Moreover, for the table to be balanced, the variables need to be expressed in terms of money, and the coefficients therefore represent value relationships. However, they may sometimes be treated as if also representing physical quantities; and subsidiary physical relationships may readily be added.

Input-output representations have typically been incorporated into larger modeling frameworks and have been made "dynamic" by being hybridized with other modeling techniques. The most frequent application to date has been automatically to link some econometrically determined variables with an I-O matrix. This linkage has taken place mostly in the modeling of mineral demands and in the preparation of mineral demand projections. For a complete system one has to have a bill of final demands, derived exogenously from assumptions (or projections) of GDP. The external projections may already provide a good deal of final consumption detail. It remains to convert from the original classification to the input-output classification and, as a rule, to constant currency of the same vintage as the input-output table. For example, the transformation is a matter of applying the inverse coefficients and reaggregating by the supplying industries. The procedure can also be applied to determine the differential impact of changes in the level of final demand for particular mineral or energy products on changes in the level of output of particular end uses. There are technical flaws in the approach if the industries chosen for the simulation are not final enough, but the errors involved are generally not important.

One of the simplest of these mineral models was that designed by the U. S. Federal Emergency Management Agency (FEMA) to explain how the future consumption of ferrous and nonferrous metals can be forecast by combining projections of material consumption ratios with an input-output forecast for individual industries. Components of the modeling structure were given in the flow diagram of Figure 3. They include an estimate of GNP derived from a macro forecasting model, an input-output table that converts the economic estimates to industry production estimates, and the estimated ratios explaining materials consumption per unit of industrial output for individual industries.

This application as reported by Kruegar (1976) employed an input-output table derived from one prepared by the Bureau of Economic Analysis of the U.S. Department of Commerce. Interindustry transactions are represented in dollars, though the disaggregation is small, with some 80 industries appearing in the actual table. The sum of final demands equals the sum of value added and total GNP expenditures, and the gross inputs for any industry are equal to the gross output of that industry.

The methodology employed first disaggregates GNP estimates by industry to form a final demand vector F. This requires utilizing a specially designed table embodying "demand impact transformation". Total direct plus indirect output requirements derived according to the relation

$$Q = (I - A)^{-1} F \qquad\qquad (7.9)$$

are then employed to estimate mineral consumption by computing mineral consumption ratios (MCR). Historical MCRs have been calculated from past levels of mineral consumption and industry output for individual industries. Forecasts of these ratios MCR_i^*, together with the forecasts of the corresponding industry outputs Q_i^*, provide the mineral consumption forecasts C_i^*.

$$C_i^* = MCR_i^* \cdot Q_i^* \qquad (7.10)$$

Summing over all industries yields total consumption for each mineral. Thus far, the methodology has been applied to the aluminum, chromite, lead, manganese, palladium, silver, tin, tungsten, vanadium, and zinc industries.

A much larger input-output model for the United States was similarly applied to determine mineral requirement and pollutant emissions in a project of Resources for the Future, e.g. see Ridker and Watson (1980). EPA's SEAS model which has as its core one of the earlier versions of the INFORUM (dynamic I-O) model developed by Almon et al. (1974) was modified for this purpose. Both SEAS and INFORUM incorporate procedures (mainly consumption functions) for moving the input-output matrix automatically forward in time, but this refers mainly to the mode of arriving at the details of final demand (including as one step the derivation of various price implications from the input-output solution). In this case assumptions have been made independent of the model for the rate of growth of GNP.

Based on a 185-sector input-output matrix, the bill of final demands is generated automatically, subject to a limited number of exogenous specifications. Most of the final demands are derived through regression estimates for categories of personal consumption, linked to the input-output categories by a "bridge" matrix. There is provision, however, for judgmental overriding of the automatic determinations. There is a special matrix to convert residential and public construction into the input-output categories, and another one to convert the initially limited number of categories of government expenditures into the required array of input-output sectors. There is also a capital coefficient matrix to convert levels of both normal and pollution abatement investment into corresponding current flow purchases from each of the input-output sectors. The actual computation of the demand for minerals can be expressed in dollars or physical quantities. Mineral detail in this respect includes a number of side-equations for mining and drilling, for cement and stone, for ferrous metal processing, and for nonferrous metal processing.

An extension of this approach to the international level can be seen in the Leontief et al. (1983) modeling effort. The resulting minerals projections feature production as well as consumption variables and are based on a systematic integration of the factors which determine domestic production, such as the level of final demand, import dependence, recycling rates and materials substitution. The consumption/production projections are generated for a number of nonfuel minerals. Consumption includes final demand categories, exports, imports, and changes in inventories. Production includes mine output, by-product output, imports, and releases from government stockpiles. Using technological updating for the 1972 I-O coefficients, mineral projections were prepared to the year 2000 for the United States and to the year 2030 on a global basis.

Projections and simulations of the foregoing types are the most common applications of input-output tables for mineral analysis, but they are not the only kinds. One type of application, which uses input-output more as an analytical tool than as a model, is to determine the relationship between mineral or energy consumption in aggregative form and their various macroeconomic and demographic determinants. For example, Rose and Kolk (1983) employed a particular growth rate for a region to forecast the demands for natural gas. Input-output tables are particularly valuable for this purpose, since they necessarily account for the whole array of inputs into each consuming sector and thus produce aggregative data for minerals in general, as well as for major classes of minerals, at various stages of extraction and processing. By also permitting the determination of the direct and indirect mineral requirements related to any general category of demand (e.g., investment, durable goods) or of economic activity (e.g., construction, transportation), they make it possible to expose relationships that are not easily ascertained through any other statistical system. They are particularly useful in comparing differences in the intensities of mineral or energy consumption for various ultimate purposes at different times and among different countries.

Input-output analysis can also be used to determine the impacts of mineral and energy industry investments on the economic growth of regions or countries. Examples of studies which evaluate industry direct and indirect multiplier impacts as well as backward and forward interindustry linkages are numerous, i.e. see Rose and Miernyk (1987). Specific applications that measure the impact of mineral industry development on economic growth include the Opyrchal and Wang (1981) study of the Florida phosphate industry and the Swisko (1989) study of the Arizona and Nevada mineral economies. More recently, Bocoum and Labys (1993) have modeled the export impact of the mineral processing of copper and phosphate on the Moroccan and Zambian economies.

An application for which input-output analysis is *not* suitable is determining the interrelationships among the demands for different minerals or fuels. While these commodities are joint inputs into any given product, there is no way of determining from the input-output table the extent to which they are complementary or substitutes, and therefore, the direction or size of change in any one such input that is implied by any given change in another.

8 ECONOMY INTERACTION MODELS

There has been a growing tendency among mineral and energy modelers to use two or more modeling methodologies in combination. In particular, as the modeling process has become more sophisticated, analysts have begun to borrow skills from one another, so that the methodologies which have emerged are more difficult to classify. This phenomenon has risen not only because of the need to couple economic and engineering considerations, but also because of the interrelationships that exist between the mineral and energy sector, the environment, and the growth of national economies. Although this integration has taken a variety of forms, it usually proceeds by model hybridization or model linkage.

Hybridization basically involves constructing models that combine two or more modeling methodologies. For example, it is not uncommon for the demand side of an energy model to be constructed econometrically, while the supply side might be principally engineering in character. In such cases the output of several different models might be needed to provide a comprehensive analysis of a particular energy problem. Linkage involves the coupling together of two or more models in a systematic fashion. For example, an input-output model of the energy sector can be linked with an econometric model of the macroeconomy. Analyzing feedback effects is important in this case, such as the measuring impact of energy prices on economic growth. The expansion of this modeling activity can be witnessed in the different forms that have emerged.

Economy interaction models, however, have also evolved as an independent form of macro-economic-energy interaction models, whose methodological structure goes far behind simple hybridization. Such models can represent this interaction as either aggregated or disaggregated economic equilibrium. The method behind such models can be econometric, computable general equilibrium, input-output or parametric. Such models can describe interactions in a single region or country as well as in several countries such as the European Union, or globally in the world.

INTEGRATED ENERGY MODELS

Early integrated energy models were largely energy specific. The models to be combined have tended to be of a programming nature, often also linked to some form of econometric model. Computable general equilibrium models of this type also exist. One useful effort to develop a methodology to deal with hybrid energy models was the combined energy model approach of Hogan and Weyant (1980). A theoretical, as well as a computational, approach was employed for reaching equilibrium solutions with a combined set of models. The previously most well-known application of this approach was the Project/Independence Energy Evaluation System (PIES) constructed for the EIA (1979). This model includes a macroeconomic model, an econometric demand model, and a programming model explaining fuel supplies, conversion and shipments.

Another early development of integrated models involved the linking of input-output models of the energy sector with macroeconometric models. For example, the Hudson and Jorgenson (1974) model consists of a macroeconometric growth model of the U.S. economy integrated with an interindustry energy model. The growth model consists of submodels of the household and producing sectors, with the government and foreign sectors taken to be exogenous, and it determines the levels and distribution of output valued in constant and current dollars. The model determines the demand for consumption and investment goods, the supplies of capital and labor necessary to produce this level of output, and the relative equilibrium prices of goods and factors. The model is dynamic and has links between investment and changes in capital stocks and between capital service prices and changes in prices of investment goods.

The model is then linked to an interindustry energy model by estimating the demand for consumption and investment goods together with the relative prices of capital and labor. The interindustry model employed is based on a nine-sector classification of industrial activity. Production submodels are developed for each sector. These submodels treat as exogenous the prices of capital and labor services determined in the growth model and the prices of competitive imports; and for each sector they determine simultaneously the sector output prices and the input-output coefficients.

The sector output prices and the demand for consumption goods from the growth model are used as inputs to a model of consumer behavior that determines the distribution of total consumer demand to the nine producing sectors. The distribution of private investment (government and foreign) is determined exogenously, and it completes the final demand portion of the model. Given final demands, the input-output coefficients can be used to determine the industry production levels required to support a given level and distribution of real demand.

The Hudson and Jorgenson model has been used to forecast long-term developments in energy markets within the framework of consistent forecasts of macroeconomic and interindustry activity. The model has also been used to analyze the impact on energy demand of alternative tax policies, including a uniform BTU tax, a uniform energy sales tax, and a sales tax on petroleum products.

Among other integrated models, Groncki and Marcuse (1980) are responsible for the BESOM model which combined a programming model of energy supply with a long run macroeconomic growth and input-output models. One example of the use of this framework has been to analyze the economic impacts of fuel scarcities. Constraints can be placed on the availability of fuels and resources, and the required fuel substitutions are determined. Coefficients in the I-O model are revised to reflect the new fuel mix, and the I-O model is again solved with the revised mix. Iterations are required between the two models in order to obtain a solution in which the energy demands and fuel mix are consistent in the two models. Impacts on the macroeconomic growth of various industrial sectors in the economy can then be evaluated.

As a final example, the INFORUM Long-term Interindustry Forecasting Model is an input-output model developed in the University of Maryland's project for interindustry forecasting, i.e. see Almon et al. (1974). The model determines constant-dollar output for 78 sectors of the U.S. economy, consistent with given levels of final demands. These final demands are forecasts of INFORUM's macroeconomic model. Inputs to the macromodel include forecasts of population growth, government expenditure, money supply, and mineral and energy prices. The INFORUM model also estimates fuel use by industry. Unlike models that represent only the industrial sector, INFORUM ensures complete supply and demand balance in energy use.

ENERGY-ECONOMY MODELS

These kinds of model go beyond the integrated input-output models to have a character of their own. Such a large number of these models have been constructed for different regions and countries that an exhaustive review is not possible here. Reviews assembled by other authors deal with particular methodologies, particular frameworks or particular regions.

Munasinghe (1990), for example, considers the large, multisector models in an energy planning context. Such models were constructed either as additions to already existing general purpose macroeconomic models, or they were specifically constructed to analyze energy supply and demand relationships. Such models consist of a number of energy variables with various driving variables. A simple model, for example, would relate how total energy use and prices result from a driving force on the demand side and production variables and constraints on the supply side. More detailed models typically distinguish between the various end-uses of energy (heating, processing, transportation, etc.) and energy sources, (natural gas, hydro, coal, petroleum, wood) each with its own production constraint.

The major sources for these empirical estimates are technological information relating to past and future production and utilization possibilities, and econometric estimates of past demand, supply, and other economic variables (production by activity sector, income, number of vehicles, rate of electrification, etc.). Much of this information consists of physical quantities of inputs and outputs, sometimes aggregated into cost estimates. If the models are to yield useful estimates, the number of independent relations and of binding constraints must at any time equal the number of energy variables projected.

Figure 4 provides a summary view of the characteristics of a number of energy models that were developed for the United States. Three groups of driving variables are identified in these models: (1) policy variables, (2) realization variables, and (3) blends of the two. In order to better distinguish between the these models, Beaver (1993) divides them according to whether they are disaggregate or aggregate economic equilibrium models. The former type of models represent domestic markets for various energy commodities, labor and capital, as well as for intermediate materials and other goods and services. The agents in the models including households, producers and governments are described as reacting to prices, such that a price is found which equilibrates supply and demand in each market over time. Because the production, investment and consumption decisions of the agents in the economy are determined within the model, the rate of growth of aggregate economic output or GDP in each region is entirely endogenous, given the models technology, preference parameters and resource and policy constraints.

Sectoral disaggregation also allows a look at the impacts in other industrial sectors outside the energy sector due to fuel constraints. Moreover, international trade in manufactured goods is another avenue for

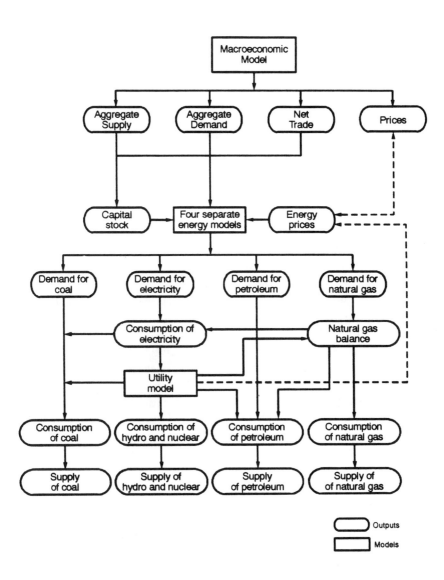

Figure 4

ENERGY-ECONOMY INTERACTION MODEL

analyzing the impacts of increases in energy prices. Finally, because these models also include a governmental sector and tax structures, they can be used to analyze the effect of various means of recycling the revenues generated by carbon taxes or carbon rights auctions.

Most of these models do not describe any non-energy sectors other than the primary factor markets (capital and labor). They also have markets for non-electric and electric energy, in which a variety of energy technologies and fuels compete for market share. Each models agents interact so that supply and demand are balanced in each domestic and international market over time, with gross domestic output in each region modeled by an economy-wide production function in primary commodities. The consumers choose their investment and consumption levels to maximize the value of their consumption streams· over time; therefore the rate of economic growth is again determined entirely within the model given the resource and policy constraints and technology and preference parameters.

These models have a richer representation of energy supply technologies, moreover, and so are able to study how a policy's cost is affected by different assumptions about the costs and availabilities of efficient and carbon free energy supply technologies. Finally, the models have a transparent representation of the rate of aggregate energy-efficiency improvement unrelated to energy price increases, so the effect of this uncertainty on a policy's costs can easily be assessed.

A more macroeconomic view of these interaction models appears in Sterner (1992) in his collection of papers summarizing a variety of energy-economy models. Other reviews appear in Capros et.al. (1992) dealing with European Union models such as HERMES (e.g. dAlcontara and Italianer, 1982), MIDAS (e.g. Capros, et.al., 1990), MEDEE (e.g. Capros, 1986) and EFOM. Reference also is made to earlier energy-economy model evaluations by Beaver (1993), by Hitch (1977) and by Samoulidis and Mitropoulos (1982).

Other forms of energy-economy models have dealt with energy planning, more particularly energy systems planning in developing countries. Munasinghe and Meier (1993) and Munasinghe (1980) have helped to develop this approach, mainly in response to a number of strategic requirements. First, the need exists for greater coordination between energy supply and demand options, and for the more effective use of demand management and conservation. Second, energy-macroeconomic links should be explored more systematically. Third, it is important to take a more disaggregate analysis of both supply and demand within the energy sector, so as to create greater opportunities for interfuel substitution (especially away from oil). Fourth, greater emphasis use can be made of the analytical and modeling tools required for energy subsector planning. Fifth, in the developing countries, economic principles can provide a more substantive input to development planning, including the techniques of shadow pricing. And finally, heightened environmental concerns can lead to greater interest in studying energy-environmental interactions.

A model planning framework developed in this context also helps to guide planners beyond the actual modeling mechanism, to resolve conflicts among the many different goals that exist in developing countries. Some goal examples include: (1) determining the detailed energy needs of the economy and meeting them to achieve growth and development targets, (2) choosing the mix of energy sources needed to meet future energy requirements at the lowest costs, (3) minimizing unemployment, (4) conserving energy resources and eliminating wasteful consumption, (5) diversifying supply and reducing dependence on foreign sources, (6) meeting national security and defense requirements, (7) supplying the basic energy needs of the poor, (8) saving scarce foreign exchange, (9) identifying specific energy demand and supply measures to contribute to possible priority development of special regions or sectors of the economy, (10) raising sufficient revenues from energy sales to finance energy sector development, (11) attaining price stability, and (12) preserving the environment.

Energy-economy models typically recognize the global consequences of energy use. Individual country descriptions can be combined in an international economic and environmental matrix. Thus both the world economy (through trade and financial linkages) and the natural resource base (through interactions involving mineral and energy resource depletion or global climate change) can impose exogenous constraints and conditions on national-level decision-making. The energy sector becomes part of the whole economy. Therefore, energy planning requires analysis of the links between the energy sector and the rest of the economy. Such links include the input requirements of the energy sector, such as capital, labor, minerals, and environmental resources as well as energy outputs, such as electricity, petroleum products, woodfuel, and so on, and the impact on the economy of policies concerning availability, prices, taxes, etc. Applications of these models particularly using one of the above optimization or linear programming methodologies together with a hierarchical approach can be found in studies by Meirer (1985a, 1985b, 1986), Meirer and Mubayi (1981), Munasinghe and Meirer (1985), and Munasinghe (1986, 1988, 1990).

Another area of energy-economy modeling has been the large-scale international models which examine trade between major regions of the world and attempt to include energy production and consumption within that model. Scenarios are generated with computer simulation models which explore the impact on international trade and growth of changes in energy conditions. The latter include rising oil prices as well as increased oil depletion and energy conservation. Some of these models include energy as only a minor sector of the whole model, while others consider energy as the driving sector of the model. Examples of these include the Hughes and Mesarovic (1978) world integrated model (WIM), the Herrera and Scolnik (1976) version of the Bariloche Model, the Linneman (1976) MOIRA model, and the IIASA sets of models described by Basile (1979). More recently, Klein (1992) discusses several similar global models, Kim and Kae (1993) report on the application of an integrated energy policy model for Korea, and Harvie and Thaha (1994) have modeled the impact of oil production on the Indonesian economy.

Some of the research on modeling these international interactions has taken a trade theory approach. To begin with, a class of abstract models exist based on trade and balance of payments theory such as that of Chichilnisky (1981). They typically feature a two-sector balance of payments model which analyzes the relations between oil prices and output, employment and prices of goods in industrial economies. The industrial or developed country region is a competitive market economy that produces two goods (consumption and industrial goods) with three inputs (capital, labor and oil). It trades industrial goods for oil with a developing oil exporting region which acts as a monopolist. The general equilibrium solution of such models determines endogenously the principal variables of the industrial region: output and prices of industrial goods as well as their factor employment and their prices. Simulations of the model can determine the impact of different levels of oil prices on the two regions.

ENERGY-ENVIRONMENT-ECONOMY MODELS

A recently growing use of integrated-energy models has been to evaluate the impacts of energy demands on the environment and to include the economic costs of related pollution abatement. Mention has already been made of integrated input-output modeling. Not only are the studies of Ridker and Watson (1980) interesting, but also those of Rose (1974, 1977, 1983). More recent modeling efforts have addressed the problem of potential worldwide warming due to increasing accumulation of greenhouse gases such as carbon dioxide and methane in the atmosphere (Ruth, 1993, 1995a,b). A less active area for modeling has been the problems of high altitude ozone depletion due to excessive releases of chlorofluorocarbons used mainly in refrigeration devices, pollution of the oceanic and marine environment by oil spills and other wastes, and overdepletion of fresh water, animal and mineral resources. The structure, applications, and validity of models of growing CO_2 emissions and how the greenhouse effect and climatic change can be minimized have recently been evaluated by Hope (1993) in a collection of papers from the 1992 conference of the International Association of Energy Economists. Econometric models concentrating on emissions and global temperature change have been constructed by Cohen and Labys (1994, 1996, 1997). Problems of mining and the environment are reported in Eggert (1994).

According to Grubb (1993), many models which have attempted to compare costs of abatement against the expected environmental benefits need further development. For example, expected environmental benefits are unquantifiable at present because of inadequate understanding of both climate systems and the likely economic responses to climate change. Much of the knowledge required for estimating the costs of abatement over a 10-20 year period now exists, but only in disparate modeling studies, e.g. see Manne and Richels (1992) and Wirl (1995). The most important and uncertain long-term factor is technological development and, in particular, the capturing of the risks and uncertainties involved in both climatic and energy systems.

So far, such difficulties have resulted in modeling efforts of only limited success. However, Dowlatabadi and Morgan (1993) have suggested new integrated approaches. Examples of such models which have already been constructed appear in Peck and Tiesberg (1993), Hinchy et. al. (1994), Eden (1993), Hourcade (1993), Barker et al. (1993), Suwala et al. (1997), and Manne et al. (1995). Further reviews and evaluations of such models can be found in Beaver (1993), Rose and Tolmasquin (1993) and Wilson and Swisher (1993). Models which concentrate particularly on carbon tax impacts include Farzin (1995, 1996) and Farzin and Tahvoren (1996).

COMPUTABLE GENERAL EQUILIBRIUM MODELS

This modeling development would replace the idea of several integrated models with one in which all of the sectors together within the economy in aggregate constitute a single or unique model. Based on the Walrasian tradition, applied general equilibrium models (CGE) describe the allocation of resources in a market economy, as the result of the interaction of overall demand and supply, leading to equilibrium prices. As explained by Borges (1986), the building blocks of these models are equations representing the behavior of the relevant economic agents: consumers, producers, government, etc. All of these agents require and produce goods, services and factors of production, and mineral and energy commodities, all as a function of their prices. The models assume that market forces will lead to equilibrium between supply and demand, and accordingly, prices that clear markets, allocate resources, and distribute incomes. In this respect the CGE methodology could apply to many of the energy-economic modeling frameworks described above.

Attempts to analyze mineral and energy markets and industries using this approach typically analyze these sectors as a partial component of a more general system. One CGE model which links the mineral sector to the U.S. economy is that of Lin (1991). In the case of the interaction of the energy sector with the U.S. economy, applications exist in the models of Borges and Lawrence (1984) and Manne (1977). Research has also taken place on integrating a more general commodity model with the Australian economy by Hanslow (1993) and Higgs (1987). A more relevant approach for Eastern Europe is the model developed by Blitzer and Eckaus (1985) which projects energy demand in national economies with a changing structure. Two-way feedbacks occur between the costs of meeting energy demands and of financing new energy supplies, and the economy as a whole. Energy demand is viewed as it affects aggregate economic growth, the balance of payments, and sectoral patterns of growth.

9 RESOURCE BALANCE MODELS

The analysis and modeling of the overall energy system including supply and demand sectors as well as all fuels and energy forms has been stimulated largely by the need to develop forecasts of total energy demand. This approach often constitutes less of rigorous modeling than one of attempting to provide sectoral and accounting balances. Supply-demand balancing consists basically of assigning specific energy sources to corresponding uses. Examination of past and present energy balances allows the energy analyst to determine the evolution of supply and demand within a comprehensive framework, the bottlenecks that exist, and how supply and demand has adjusted to constraints. Forecasting requires that supply-demand balances must be developed for future years. Projected energy shortages and surpluses by fuel type and usage category must be reconciled, for example, by increasing or decreasing energy imports or exports, inter-fuel substitution (where technically and socially feasible), augmenting domestic conventional and non-conventional energy sources, reducing demand through pricing, rationing and physical controls. Thus, some of these equilibrating measures will require going back to earlier steps to re-adjust the supply and demand analyses and forecasts.

Much of the initial work in this area involved the development of overall energy balances for the United States in which forecasts for individual fuels were assembled. These forecasts highlighted many problems involving factors such as resource definition and interfuel substitution, which must be handled in a consistent manner for all fuel types and sectors. In order to produce these forecasts, a methodology had to be developed using the energy balance concept. However, the methodology never advanced to the point of formal quantitative modeling. It has been included here only to complete the modeling framework. Most typically these models have been separated into energy analysis models and energy gap models.

ENERGY ANALYSIS MODELS

These models possess less of a formal structure than that of input-output models. According to Ulph and Folie (1977-1978), they consist of two types: (1) accounting models which attempt to give a comprehensive assessment of the energy costs involved in producing different products, and (2) net energy models which examine the ratio of energy inputs to energy outputs associated with certain processes, usually energy or food production. These methodologies are based on data which provide detailed accounts of flows of energy. As such, they could provide a basis for more formal models of the type described above.

As an accounting approach, the energy balance system focuses attention on a complete accounting of energy flows from original supply sources through conversion processes to end use demands. The approach accounts for the intermediate consumption of energy during conversion processes as well as efficiencies at various points in the energy supply system. An important characteristic of the system is that prices are not determined by the market forces of demand and supply. Rather, prices are determined by the interaction of both consumer preferences and technological considerations, the latter reflecting the sum of the direct and indirect energy required to produce any commodity. Any resulting analysis assumes that energy is the only nonproduced input to any production process, which involves either ignoring other inputs such as labor, or attempting to aggregate all nonproduced inputs into a single factor called energy. Obviously such analysis ignores important differences in the nature of different inputs, the finiteness of resources, and the allocation of resources over time.

An earlier review of these kinds of models can be found in Hoffman and Wood (1976). One of the first systematic attempts to account for all energy flows in a consistent manner was that of Barnett (1950). Barnett's approach involves obtaining a national energy balance of energy supplies and demands by type. The emphasis was on quantity flows expressed in physical units and a common unit, the BTU. This approach was earlier extended and refined by Morrison and Readling (1968). Some examples of international energy applications can be seen in works by Heuttner (1976) and by Webb and Pearce (1975).

ENERGY GAP MODELS

These models may even have less of a technological or structural framework than energy analysis models. According to Gately (1979), they typically contain no explicit functional relationships among the variables, nor any equilibrating mechanism to ensure consistency among supplies and demands in the various markets. For any particular set of assumptions about the underlying political-economic environment, there are separate estimates of demand and supply which are disaggregated by region and fuel type. The demand estimates, for example, are developed by relating demand to aggregate economic activity and trends in energy consumption. Independent estimates of supply of major energy types are developed and compared with the demand estimates. Differences are resolved, usually in a judgmental way, by assuming that one energy type is available to fill any gap that may exist between supply and demand. This energy type is normally assumed to be imported petroleum, including crude oil and refined petroleum products. The DuPree and West (1972) study provides an example of the execution of a forecast employing this methodology. Other examples include those of National Petroleum Council (1974), Exxon (1977), OECD (1977), U.S. Central Intelligence Agency (1977), and the Workshop on Alternative Energy Strategies (1977).

10 TRANSITION MODELS

Most of the above modeling methods have considered the underlying mineral and energy market structures to comply with conventional market paradigms such as competition, monopoly or oligopoly. However, none of these structures specifically reflects that of a planned economy. What should be recognized is that a planned commodity model might conform to what has been explained as a controlled market model. Or more realistically, a planned commodity model might reflect the interindustry requirements specified or forecast using input-output analysis. The latter method was used frequently in modeling industrial activity in the former U.S.S.R. and East European economies.

As the markets in these economies have opened to competitive behavior, the need to model markets during this transition period has required investigating modeling approaches which are applicable in this context. Such changes can normally be modeled without undue difficulty. To begin with, mineral or energy markets which undergo transition can now be described as a competitive equilibrium model in which demand and supply variables interact simultaneously with price variables to represent market clearing. In the case of spatial allocation modeling, the allocation of mineral and energy commodities over space which was obtained through cost minimization of the objective function now can be opened to market and price adjustment. This might simply require employing the net social payoff or quadratic programming approach so that quantities and prices interact simultaneously through the introduction of demand and supply equations, e.g. see Suwala and Labys (1997).

In most cases the variables now represented in either model form adjust from being planned or fixed to now being flexible. This phenomena occurs for capital or capacity, energy and materials, and labor. In most cases the prices of mineral and energy outputs as well as factor inputs have to rise to world price levels. Externalities such as environmental factors also begin to enter the consumption and production functions.

The problem of modeling mineral and energy markets in transition, therefore, is one of employing methods which reflect a shift from planned towards competitive market equilibrium and disequilibrium adjustments. While this would appear to be a relatively straightforward procedure, the more difficult problem is that of finding sufficient data of adequate quality which can provide the basis for quantifying the variables and of constructing the equations in the designed model structures.

11 EVALUATION OF MODELS

Model evaluation helps to determine how models can be used to improve the effectiveness of mineral and energy policy analysis. The need for evaluation and better communication can be illustrated by the study of Greenberger and his associates (1983) who have surveyed the views of energy experts and others regarding the quality, the attention received, and the influence of energy studies during the late 1970s. Their study suggested that while attention received and influence were positively correlated, quality tended to be negatively correlated with both attention received and influence. The study offers no explanation for the latter, but throughout emphasizes the importance of the evaluation and the effective communication of the modeling results.

These elements of mineral and energy model evaluation have been much discussed and debated, i.e. see Gass (1977), Greenberger, et al. (1976) and Labys (1982). These researchers distinguish between two fundamental aspects of model evaluation: validation and verification. Structural validation refers to the correspondence of a model to the underlying processes being modeled and is based on: (1) the conceptual specification of the model, (2) the specification and application of the measurement process by which the model data are generated or obtained, (3) the specification and analysis of the scientific hypotheses derived from theory underlying the model and to be tested with the model data, and (4) the selection of the final model best supported by the scientific laws, principles, maintained hypotheses, and tested hypotheses which have emerged from the research process. Validation of a model structure starts with the replication of measurements and hypothesis testing, but also includes analysis and/or counter-analysis involving the variables and concepts integral to the policy issues for which the model was intended.

Content validity, is usually singled out from structural validity for the latter purpose. Both policy evaluation and policy analysis require that models reflect the appropriate policy concepts and instruments. A policy evaluation model will be simpler than a policy analysis model in this regard in that only the policy actually implemented and being evaluated must be included. Policy analysis models are more complicated in that the policy

instruments and concepts suitable for the alternative policies of potential
interest and importance to the various constituencies concerned with the
issue(s) of interest must be included. Further the model must be explicit
concerning the resolution of "facts" and/or value judgments, which are in
dispute among the various constituencies. Crissey (1975) has made the
evaluation of such "facts" or contention points a central feature of his
approach to policy model analysis.

Predictive validation determines if the scientific information and results
included in the model are sufficient to discriminate among the future
policies being considered. If the range of scientific uncertainty spans the
range of policy dispute, then the model's usefulness in policy research is
very limited.   Model-based studies may sometime only report point
predictions and not information on prediction confidence limits or
sensitivity analysis of prediction to changes in input data and/or structural
coefficients, consistent with known or conjected uncertainties in the
underlying measurement processes and scientific results. The lack of this
additional information may suggest an unjustified precision of analysis. For
example, Shlyakhter (1994) has shown how trend and errors in past energy
forecasts can be used to model uncertainties with related variables.
Analysis of predictive power is thus an important aspect of policy model
analysis quite independent of the structural validity of the components of
the model.

Closely related to the various dimensions of model validity is the
validity of the data associated with the model. Data validation must include
not only evaluation of the measurement process by which the data
component of model structure is developed, but also the process by which
the data required for model applications are obtained. While data and
measurement process evaluation are closely related to model evaluation,
particularly to that of model structural and predictive capability, it is
probably useful to single out this aspect of validation since it typically
receives too little attention in policy modeling and research.

Crissey (1975) has a similar perspective on the elements of policy model
validation. He has emphasized that the credibility and utility of a policy
model will depend upon its treatment of the factual, behavioral, evaluation,
and structural issues in dispute. Disputed issues should be represented in
the model in a manner facilitating analysis of alternative resolutions. Such
issues comprise the model's contention points. According to Crissey, a
contention point is said to be critical if changes in its resolution significantly
affect the model conclusions, and is a *contingency point* if changing the
resolution of this contention point in combination with others results in a
significant change in the model result (Crissey, 1975; pp. 83-88). This

concept of model contention points provides a useful focus for structural, content, and predictive validation.

In contrast to validation, mineral and energy policy model verification refers to the evaluation of the actual model implementation. That is, does the final quantitative model structure conform to what the modeler intended? Verification is thus more mechanical and definitive than model and data validation. Gass (1977) has suggested that policy model verification is the responsibility of the modeler, and that evaluation should be limited to review of the verification process.

A final aspect of policy model evaluation concerns usability. This dimension of evaluation refers to both the sufficiency of documentation to support model understanding and applications, and the efficiency of the overall system. The technical documentation and materials sufficient to inform potential users of the nature of a model's structure, content, and predictive characteristics, as well as to support interpretation of model-based results, are essential for any policy model. The need for documentation to support independent application of a model, including user guides, system guides, and test problems will depend upon the model application environment. Of course, even if the intent is for the modeler to conduct all applications, there still should be evidence that application procedures have been developed, and that a reasonable applications practice is in effect.

The elements of model evaluation serve as a guideline for self-evaluation by the modeler, and for independent model evaluations. Given these guidelines, what can be said about the process of independent evaluation, an activity of increasing importance in establishing the credibility of policy models. The MIT Model Assessment Program has identified several such elements as important for independent evaluation: (1) Review of literature, (2) Overview assessment, (3) Independent audit, and (4) In-depth assessment, i.e. see Labys and Wood (1985). The major distinction between the approaches concerns the information used in evaluation.

A summary of the relationships between these evaluation steps is given in Figure 5. A review of the literature for a model, or set of similar models, focuses upon model formulation, measurement and estimation issues relating to model structure, applicability for analysis of specific policy issues, and so on. Such a review may be both descriptive and evaluative. An example is the review by Taylor (1975) of electricity demand models comparing model structure with an "ideal" structure. In its various forms, literature review and analysis is the traditional means of model analysis. Issues of approach, logic, measurement and interpretation are formulate

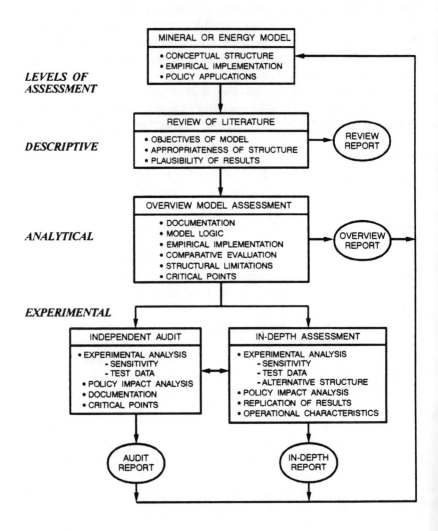

Figure 5

APPROACHES TO MODEL ASSESSMENT

and analyzed.   Issues of actual implementation are less susceptible to analysis with this approach.

An overview assessment examines the underlying technical model documentation, especially the computer code, for a more precise analysis of the model's structure and implementation.   An overview evaluation can identify a policy model's critical points, but it will only occasionally be able to pass judgment on the adequacy of the model's treatment of them.   It is important to quantify the nature and level of a model's explanatory and forecast errors, as explained in Labys (1982).   The assessment of a model's validity and applicability generally requires the acquisition and analysis of experimental data.

An independent audit evaluates a model's behavior by analyzing data derived from experiments that are designed by the assessors but run by the model builders.   An important element of the procedure is that the assessment group is "looking over the model builder's shoulder" while the experimental runs are being made.   This is essential to the accurate interpretation of the results produced by the experiment.   An audit report should use the experimental data together with the analytical material developed in previous stages of the evaluation process to determine a model's validity in as many key areas (critical points) as possible.   Audit procedures have the advantages of being relatively quick and inexpensive. With complex models, however, there will generally be some critical points that cannot be fully evaluated through an audit.

An in-depth assessment develops experimental data through direct, hands-on operation of a model.   Direct operation makes it feasible to carry out more complex tests, particularly when the tests require modification in model structure rather than single changes in model parameters and/or data. Because of the significant costs of in-depth evaluation, it is probably most efficient to conduct exploratory analysis through an independent audit before embarking on more detailed evaluation. After an in-depth evaluation has been completed, audits might subsequently be used to update the evaluation as new versions of the model are developed.

Rather comprehensive attempts at model assessment have continued under the auspices of the Energy Modeling Forum at Stanford University. Of recent interest has been the validation of existing important world crude oil models. For example, Kress, Robinson and Ellis (1992) and Beaver (1993) have reported on the evaluation of the structures of the important world oil models on a comparative basis. A more formal estimation of the econometric response surface of aggregate oil demand in the OECD countries was reported by Huntington (1993) based on nine different world oil models.   The response estimates were based upon scenario results reported for the 1989-2010 period in the Kress et.al. (1992) study. The response surface approach provides a parsimonious summary of model responses. It enables one to estimate long-run price elasticities directly rather than to infer such responses from 20-year cross-scenario results. It also shows more directly the significant effect of initial demand conditions (in 1993) on future oil demand growth. Due to the dynamic nature of the oil demand response, past prices were shown to exert a strongly positive effect on future oil demand in some models, but little or even a negative effect in other models. The results of the study suggest that demand modelers should be much more explicit about what their systems reveal about the extent of disequilibrium embedded in the starting conditions of their model.

We have briefly surveyed the elements of model validation and verification, and approaches to independent evaluation. It should be emphasized, however, that a modeler has the primary responsibility for designing, conducting, and reporting the results of validation and verification efforts. When the modeling research is reported in refereed journals and monographs, a modeler will be guided by general scientific practice. However, modeling research intended to be used in policy analysis, and often not submitted for refereed publication, places an additional burden on a modeler. In addition to providing documentation consistent with "good scientific practice", a modeler must also satisfy the information needs of a (usually) less technically oriented audience interested primarily in model based results, not the modeling research itself.

The essential requirement is to provide these diverse mineral and energy model audiences (other modelers, policy analysts, and those affected or influenced by model based studies) with the information necessary for themselves to judge the adequacy and legitimacy of particular models. No amount of independent evaluation can substitute for a well conceived and executed plan to satisfy these legitimate information requirements. Hence the prospective modeler intending to contribute to the analysis and understanding of the current and prospective energy policy issues should be aware that good models and analysis are not sufficient to ensure the credibility and utility of model based policy studies. The findings of Greenberger and his associates, mentioned at the beginning of this section, was that for the major energy studies of the 1970's (and many relied heavily on models) perceived quality was negatively correlated with attention received and influence. This is a sobering reminder and challenge to mineral and energy modelers to consider carefully, and to satisfy, the legitimate information needs of their audiences.

12 SELECTION AMONG APPROACHES

One of the most difficult tasks in mineral and energy model implementation is deciding which single modeling approach or which combination or hybridization of methodologies will best solve the modeling problem at hand. It is important to realize that there is no such thing as an "all-purpose" model. Different model structures lead to the solution of different kinds of mineral and energy policy problems. At the same time, however, some relation does exist between the requirements of model specification and the methodologies employed. Specifications that require temporal, process or spatial configurations might employ econometric, engineering process, or spatial equilibrium models respectively. To help make these and other more difficult decisions of model selection, the major methodologies are now compared on the basis of their relative strengths and weaknesses. Time and space limit the evaluations to only the three more utilized approaches.

ECONOMETRIC MODELS

It should be emphasized that, although econometric models are relatively strong in specific areas, this does not mean that econometric models will always be successful in these applications. For example, the nonavailability of data at an acceptable cost may preclude the use of econometric techniques in certain circumstances. Likewise, because econometric techniques are relatively weak in certain applications, this does not mean that they cannot be used for these applications. It merely indicates that great care must be taken, and a model builder must be prepared to modify his or her approach to forecasting to take account of a technique's limitations. As shown in Table 2, econometric models are perceived as being strong at linking mineral and energy demand with macroeconomic trends. Econometrics deals with aggregated and generalized relationships of this kind rather well. On the other hand, linking mineral .

Table 2
STRENGTHS AND WEAKNESSES OF MINERAL AND ENERGY
MODELING METHODS

STRENGTHS AND WEAKNESSES OF ECONOMETRIC MODELS

Strengths	Weaknesses
Linking mineral demand with macroeconomic trends	Linking mineral demand with specific engineering decisions by mineral consumers
Short- and medium-term forecasting	Long-term forecasting
Analyzing cyclical instability	Analyzing structural change
Assessing impact of supply disruptions (strikes, cartels)	Assessing impact of demand disruptions (price control, rationing)
Projecting generalized technological and institutional change	Projecting specific technical change
Analyzing marginal reversible substitution	Analyzing radical irreversible substitution
Management of stockpiles (acquisition and disposal)	Management of conservation and substitution programs

STRENGTHS AND WEAKNESSES OF PROGRAMMING MODELS

Strengths	Weaknesses
Analyzing mineral demand in a well-defined range of specific industry applications	Analyzing diversified mineral demands in loosely defined applications
Analyzing nonmarginal substitution	Measuring price elasticity of demand
Analyzing technological change	Measuring income elasticity of demand
Long-term forecasting	
Analyzing structural change	Analyzing business cycle and other macroeconomic impacts on demand
Analyzing specific consumer decisions of process technology changes	

STRENGTHS AND WEAKNESSES OF INPUT-OUTPUT MODELS

Strengths	Weaknesses
Accounting comprehensively for all minerals	Ascertaining specific mineral
	Requirements under changed
	Assumptions or circumstances
Analyzing the effect of structural changes	Analyzing the effects of business cycles
Analyzing the impact of assumed technical or technological change	Accounting for the effects of unspecified ongoing technical change
Analyzing the effects of major macroeconomic change on mineral requirements	Analyzing the effect of minor macroeconomic changes on mineral and energy requirements
Requirements forecasting	Price forecasting
Final demand analysis	Difficulty of estimating coefficients

demands with specific engineering decisions made by narrowly defined classes of consumers is much less amendable to econometric treatment. Thus the econometric model builder must supplement his generalized statistical analysis with specific investigations of the views of mineral and energy consumers, especially where there are relatively few consumers or few end-users for a mineral.

Econometric models are strong for short- and medium-term forecasting. The basic institutional structures that they embody, both in relation to the macro-economy and to industry, tend to remain relatively stable or to move in a predictable manner. For very long term forecasting, such as a 15- to 25 year time frame, serious forecasting errors can result from major structural changes in the economy as well as in particular mineral or energy markets. When econometric models are employed for such purposes, they usually are supplemented by information from professional commodity modelers and other specialists in technological forecasting, i.e. see Labys and Pollak (1984).

Econometric models are also perceived as being relatively strong in analyzing cyclical instability where the behavior of a specific mineral or energy market can be established from an analysis of past cycles. On the other hand, an econometric model is less capable of handling structural change. A good example of this is when a market changes from being one characterized by transition from monopoly conditions and administered prices to one that is competitive. Examples of such changes would be the zinc market in the mid-1970s, the aluminum and nickel markets in the late-1970s, the molybdenum market in the early-1980s and the shift from planned to market economies in Eastern Europe in the 1990s. The willingness on the part of the econometric model builders to modify a price adjustment mechanism in the light of any perceptions of structural change in the market may be critical to the success of a model.

Econometric models are also relatively strong in assessing the impact of supply disruptions because of the central role of a market-clearing price mechanism. On the other hand, econometric models are less satisfactory for assessing the impact of demand disruptions. While supply shocks tend to change the aggregate quantity of material supplied, demand disruptions tend to change the rules under which the market place operates. These models can be supplanted to deal with problems of risk and uncertainty in mineral and energy markets by manipulating them with risk analysis software.

Econometric models tend to be strong in predicting generalized technical progress because such trends are typically well established over the historical period to which a model refers. By contrast, specific technological changes represent a break with past trends, and there is no way that an econometric technique can easily accommodate these, other than through the use of an engineering model adjunct to the econometric formulation. Clearly, considerable skill on the part of the econometric model builder is necessary to distinguish between developments that represent the trend of technical progress and those that represent a fundamental break with the past.

Because most econometric models are based on the assumption of perfect competition, they tend to be strong in analyzing marginal and reversible substitution of the classic economic kind. In contrast, radical substitution involving asymmetric or irreversible changes is much harder for a model to handle.

Finally, econometric models are relatively strong for developing policies for managing economic stockpiles, for example, such as international buffer stocks, since the objectives of the stockpile managers can be incorporated in a set of economically rational statistical relationships. On the other hand, good management of strategic or military stockpiles is much more difficult to achieve with an econometric model. While such stockpiles are governed by clear behavioral rules, these rules are not a function of economic pressures but operate irrespective of market conditions.

SPATIAL AND PROGRAMMING MODELS

These kinds of models are strongest for exploring the effects of changes in technology, policy, or economic variables on mineral and energy markets. Because of their capability for incorporating considerable engineering detail related to the production and use of materials, these models are particularly capable of analyzing the effects of changes in variables that are directly related to production costs and/or design decisions. This capability includes: (1) analyzing the effects of a change in the design of a product on the demand for specific materials (e.g., the effects of redesigning automobiles produced in the United States on the demand for materials), and (2) analyzing the effects of a new production process, such as direct casting of carbon steel sheet, on the cost of production.

The ability to incorporate engineering detail also allows us to investigate the effects of changes in such variables on substitution. These models have been used to analyze the effects of changes in factors that influence the costs or properties of finished materials on their competitive position vis-a-vis substitutes. For instance, a system dynamics copper-aluminum substitution model can provide a framework for analyzing the effects of changes in the prices of competitive materials, and the costs of fabricating, maintaining, and using them, on the total costs of producing a functional unit. This is particularly true of the linear complementarity programming model variant. The flexibility and ability of the model structure to incorporate detail also allows the analyst to include information about time lags, asymmetries, and irreversibilities associated with the substitution process within the model. This kind of detail analysis also is useful for modeling electrical utility systems.

While these models are strong in incorporating disaggregate detail, they are relatively weak in measuring parameters related to the supply of and demand for materials or energy inputs. They cannot be used to measure directly: (1) the elasticities of supply or demand, elasticities of substitution, direct or cross-price elasticities, income elasticities, or (2) the time lags associated with the independent variables that determine supply and demand. Thus the models are dependent upon other techniques, such as econometrics or engineering analysis for estimates of these parameters.

These models are also relatively weak in short- and medium-term forecasting, with one exception. The most straight forward specification of a demand function (which uses an income variable estimated by engineering analysis and which usually neglects price effects) can provide accurate short-term forecasts when the material requirements per functional unit are known, applications within an end-use sector are few, and the short-run price elasticity of demand is negligible. Other similar models are not usually used for forecasting, except in the long term. These models are relatively strong for long-term forecasting only because the task is difficult for any other type of model.

The strength of the programming approach is that it allows us to incorporate future technological change that we have good information about. Therefore, if we are forecasting the demand for copper, we can specify that at a certain point in the future the use of fiber optics technology will reduce the demand for copper in the telecommunications industry by a specified proportion. Alternatively, we could specify an engineering production function for a new production technology, such as the liquid dynamic compaction process for producing rapidly solidified metal powders, to simulate future costs of production. Of course, if the past is a guide to the future, we can be assured that there will be numerous changes in technology, as well as in other variables, that cannot easily be anticipated. In short, we cannot be confident that any of the methodologies currently at our disposal can provide useful long-term forecasts.

Another factor to be considered is model linkage. Programming models are relatively strong in linking either demand for materials with specific engineering decisions made by consumers of materials, or the costs of production with specific production technologies. This is because the models can be made as detailed or as disaggregated as one desires. On the other hand, they are less useful for linking supply/demand relationships with macroeconomic trends. This is also a result of the disaggregated nature of the models. Thus a programming model of the aluminum industry might be capable of representing the effects of decisions by aircraft manufacturers to use significant quantities of aluminum-lithium alloys in the future, but it would not be useful for relating the demand for aluminum to changes in the gross national product or the aggregate industrial production index.

There are two additional major weaknesses in programming models, both of which are correctable but which are seldom corrected. First, the demand for input materials is usually calculated simply as the requirement for the input material per unit of output. The price of the input material is implicitly assumed to be unimportant in the range of interest. This may be a good assumption in some cases, but generally it is not. There are three ways to handle this problem. One is to use engineering analysis to estimate the response of a production process to changes in the price of the input materials, as demonstrated by King and Reddy (1981). The second is to include an optimizing routine that allows a model to alter the mix of input materials, depending on relative prices. This has been used by Clark et al. (1981) to model the demand for scrap and hot metal in steel making. The third is to construct some form of STPA model in which demand and supply quantities as well as prices are determined simultaneously.

A second weakness is related to the fact that the demand function for finished materials must either be supplied exogenously or estimated simultaneously with the supply function. The latter option is seldom taken because simultaneous estimation is time-consuming and often expensive. The problems encountered when using an exogenously specified demand function are twofold: (1) there is the likelihood of an identification problem if both the supply and the demand function have some price sensitivity; and (2) an incorrect demand function is often used because there is nothing else available. If process programming has been used to estimate a supply function for primary copper in the United States, for example, it is appropriate to combine this function with the residual demand curve facing U.S. primary metal producers (netting out supply and demand in the rest of the world) to simulate market clearing prices. However, since it is difficult to derive such a residual demand function, it has been the practice in the past to estimate the demand for copper in the United States with an econometric model. This is not appropriate when demand includes the demand for imports and scrap as well as the demand for domestically produced primary copper.

The point is that unless these models are combined with proper and compatible demand functions, they are not suitable for explaining and forecasting market behavior. If they are combined with appropriate demand functions, as in the case of the STPA methodology, they are as capable of providing simulations of the dynamic paths of prices, supply and demand, as would econometric models. Thus a properly structured programming model is useful for analyzing issues similar to those that have been the province of econometric market models in the past. These issues include (1) the impact of supply disruptions; (2) analyzing cyclical instability; (3) analyzing policies on the management of stockpiles; and (4) analyzing trade policies.

Programming models are also concerned with the choice of the economic mix of resources and technologies. In order to model such a choice, it is usually necessary to assume that the firms in the industry act collectively (as if they were a single firm) to minimize cost. Thus optimization is the normative process, indicating what preferred decisions or directions an industry should take, given the assumptions made in the analysis about the form and nature of the constraints and the demand target (objective function). Since this procedure provides estimates of the lowest feasible bound to the cost of supplying final products, its major strength is that it can provide a benchmark for evaluating the actual functioning of firms or industries.

However, the process approach reflecting technologies has limited utility for forecasting or even predicting the state of the system being modeled. In formulating a programming model, it is usually necessary to assume that firms and other resource owners either minimize costs or maximize profits over a specific period of time. In reality, firms usually have other objectives, such as greater market share, employment, growth, or even short-term survival, in addition to profits. Moreover, the relative importance of the various factors, and the time frame in which decisions are made, varies from firm to firm.

Further complicating the problem is the fact that is has proven difficult to build robust programming models of imperfectly competitive markets and yet a large majority of mineral and energy markets are not competitive in the economic sense. It is the nature of most of the markets that a small group of firms or countries have a significant concentration of market power. Thus, given the nature of the problems (a large array of possible motivations among firms or countries, different time horizons, imperfect competition), it is not surprising that process optimization models have not proven to be useful for forecasting or simulating the state of a given market.

In conclusion, spatial and programming models, represent a hybridization of the econometric and the engineering process approaches.

They thus possesses some combination of the strengths and weaknesses of the separate methodologies. Among strengths, demand as well as supply elasticities can be embodied in the included econometric equations. This provides a linkage with macroeconomic activities and helps explain mineral or energy substitution. Where a temporal structure is employed, it can analyze market instability and market disruptions. Substitution analysis can take place not only in the econometric sense but also by including technological factors in the stage of process analysis that is convenient to mathematical programming formulations. Model extensions from QP to LCP and MIP have made the approach particularly adept at including mineral and energy investment decisions of a long-run nature. Among weaknesses, this approach tends to be commodity specific, thus omitting detail on the many derived demands. Structural change also cannot easily be handled, since the contained econometric equations are restricted to the particular time period of estimation. This also limits the potential to include technological change in forecasting future demands and supplies.

INPUT-OUTPUT MODELS

Before referring to Table 2, it is important to realize that an advantage of I-O for modeling is its comprehensiveness. This is the only approach in which the aggregate of all minerals (among other materials and inputs into productive activity) or energy forms is accounted for. It is thus a particularly valuable resource for analyzing total utilization and observing the mix of such utilization (at least on a monetary value basis) according to different categories of mineral or energy commodities. Input-output tables also provide a ready way of observing the separate implications for mineral and energy consumption of different kinds of final demands, different end use industries, and, in fact, any other specific sectors. The susceptibility of the I-O matrix formulation to algebraic manipulation by matrix algebra also makes possible the calculation of all indirect and direct mineral and energy requirements, including impact multipliers and interindustry linkages.

The very comprehensiveness of the system is, at the same time, a severe drawback if the modeler cannot work exclusively with published input-output tables. If it is necessary to adjust available I-O matrices for desired evaluations of different assumptions, a great deal of work has to be done in order to end up with full consistent alterations, either of the transactions matrix or of the technical coefficients. Worse still is the situation where an input-output table would have to be compiled from scratch. Even where reliable tables exist, considerable work may have to be done to expand such tables with newly acquired data, estimates, or borrowed coefficients in order to provide the detail necessary for adequate mineral or energy industry analysis. The shortcut of estimating mixes within more aggregative input-output industry sectors sacrifices the opportunity for precise calculation of indirect flows and requirements that the input-output system potentially offers.

Even the most detailed I-O tables, if used mechanically to determine the implications for mineral or energy inputs for different levels of output, suffer from problems of aggregation. An often cited example relates to mineral requirements for containers. A typically detailed U.S. table (496-order) contains sufficient industry breakdown to distinguish among metal barrels and drums, metal cans, and glass containers, but it doe not distinguish among aluminum cans, tinplate cans, and hybrids. In fact, it is not very disaggregate for any of the mineral commodities. Thus use of this table to evaluate the mineral requirements implied by growth in the consumption, say, of beverages would generate some amount of derived demand for steel and tin. But the actual metal input into beverage cans may vary between steel and aluminum.

The coefficients in any given I-O table may also suffer from the structural effects of the particular stage of a current or future business cycle. For cyclical reasons or otherwise, input-output ratios may be affected by the extent of capacity utilization; but even when this is not the case, it may be invalid to assume that they do not change with scale. Unless the tables are used to model only minimal changes, it is almost inevitable that scale differences will imply significant changes in the production process and efficiency mix. This is in addition to the technological change that takes place over time. Thus for an input-output table to be valid as a model for another time and place, various adjustments have to be made.

Since even the most homogenous of input-output sectors has some element of heterogeneity, the problem of external adjustment is inevitably complicated by the need to move back and forth between physical analysis and value composites. This affects the handling both of substitution and of technological change. Normally physical terms provide some basis for introducing such assumptions into a model, but the unit values of these physical units, as they combine to make up the sectoral value composites, are rarely ascertainable. It is necessary to fall back upon indexing assumptions, with unknown and sometimes wide degrees of hazard.

It follows from the randomness of given I-O tables in relation to the business cycle (and from the randomness in the coefficients that is to be expected from other causes as well) that the tables have serious drawbacks in model applications which predict even marginal change. However, the more nearly homogeneous the inputs and outputs linked by a coefficient, the less a particular coefficient is going to be affected by purely economic fluctuations, i.e., the closer it will come to being a purely technical coefficient, valid for limited amounts of change within a limited time period. So far as radical change is concerned, no model that is fitted to a particular period in history can be expected to be valid for other periods without appropriate adjustment.

Once an I-O matrix has been established and inverted, the cost of utilizing it for answering many kinds of mineral or energy requirements questions is quite small, far less than for most other kinds of detailed models. To determine the effect of different arrays of final demand on the consumption of any given mineral or energy fuel (assuming the product is represented by a single input-output sector or that connecting ratios have been established), it is unnecessary to run the whole model (matrix). Since the predetermined inverse coefficients can be utilized, evaluation of the implications for particular minerals or fuels of changes in just one or two final demand (or intermediate production) variables takes hardly any time at all.

The comprehensiveness of input-output formulations should not be taken to signify that they are full-equilibrium models. The base tables are in equilibrium because they represent a recording of history. At least they represent as much of an equilibrium as ever exists in fact rather than in theory. Basically, they are exclusively requirement models. There is nothing of the supply or price interaction that can be obtained only by linking the model with a feedback system. I-O models, therefore, are most often operated only in conjunction with some other modeling methodologies. Computable general equilibrium models coupled with I-O models currently provide an effective vehicle for modeling full-equilibrium systems.

It is often cited as a drawback of I-O analysis as a modeling device that it is static rather than dynamic. This can be overstated. Even if there is no attempt to provide automaticity of movement from one period to the next, the need to use successive static models to represent successive cross-sections of time is far from being a fatal defect. In fact, it may be an advantage to be able to enter deliberate assumptions and parameters arrived at through independent analysis and investigation rather than to rely upon the self-perpetuating system that may propagate growing errors. Above, dynamic versions of I-O analysis have clearly been cited.

One of the advantages of I-O analysis, in fact, is its transparency to the user. The direct relationship of each dependent variable to each independent variable is plainly laid out in the basic transactions table. The unit of measurement is the same for all variables, and the parameters are always in terms of dollars or similar equivalents in other currencies. Once explained, the meaning of the inverse (or "total input") coefficients is straight forward. Because such coefficients are derived from the empirically established direct coefficients, they are above suspicion.

As an analytical tool for evaluating the determinants of mineral or energy consumption at any given time in any given country, input-output matrices, if sufficiently disaggregated, can be of great value. If price deflation problems can be overcome, these matrices can also be very valuable for comparing differences in the determinants of mineral and energy consumption between one period and another in any given country; and, if, in addition, exchange rate problems can be overcome, they can be useful in comparing differences among countries.

But input-output matrices suffer from being point observations, not even averages of several years' running. For dynamic simulation or projection purposes they need laborious and very costly adaptation, unless confined to simulation of the effects of relatively marginal changes, without significant cyclic, structural, or technological implications. Despite their great detail, these matrices often suffer severely from problems of aggregation, and where they do not already exist, they are rarely worth developing for purposes of mineral or energy requirements analysis alone.

DATA REQUIREMENTS

Compared to the other modeling methods discussed, econometric models tend to require fewer data because they usually deal with aggregate relationships rather than composites of individual relationships. However, an important characteristic of econometric and time-series models is that aggregate relationships are estimated on the basis of past behavior. This requires time series data to be available for a considerable period in the past, but exactly how long is not always a clear-cut issue. Too brief a history will clearly impede an analyst's ability to develop statistically valid relationships. An analyst usually wants to have information on at least two or three business cycles so that he can understand how the relationship under consideration may have been influenced by events of a cyclical nature. On the other hand, most forms of econometric analysis attach similar weights to the most recent and the most distant data. To the extent that structural change may be occurring, errors may develop if the estimation period is too long.

Mineral and energy programming models require data concerning a large number of activities and various categories of costs, particularly regarding supply or production. For process models, the derivation of these data requires engineering analysis of the production processes. The extent of the analysis is determined by the level of disaggregation of the technology itself as well as the market structure. At the highest level of aggregation (i.e., aggregate process inputs at the industry level), the data are relatively easy to obtain from government and industry publications or on compact discs, although the reliability can suspect because of data collection difficulties. More detailed data (e.g., plant specific data) are very difficult to obtain because an analyst is required to sift through a variety of potential sources and then must visit production sites to get such data. This task is often complicated by the proprietary nature of the data.

In general, the data requirements for spatial equilibrium and programming models tend to be more extensive than those for econometric models in that relatively more detail is included in these models. On the other hand, engineering models have the advantage of not necessarily requiring time series data, although the dynamic application of these models requires such data.

Most input-output mineral models require some supplementation of the official or published country input-output tables. Mineral or energy demand projections or modeling rarely use more than a handful of end-use product or industry demands as the starting point. For input-output manipulations, this limited number of variables has to be restated as a detailed bill of final demands. If some amount of end-use detail is exogenously assumed, that much can be converted from, say, standard industrial classification (SIC) to input-output classifications by means of "bridge" or "crosswalk" tables that link the two classification schemes.

Regarding model output, there is a similar need to transform the total dollar (or other currency) mineral requirements generated by any given final bill of demand into quantities of minerals or energy commodities. Even the most detailed recent U.S. table, for example, distinguishes only one specific kind of ore (copper) and four kinds of refined metals (primary copper, lead, zinc, and aluminum). Iron ore is lumped with ferroalloy ores, pig iron with all the other kinds of nonferrous scrap, and so on. In other words, for many mineral modeling purposes there has to be a supplementary data set (in effect, sets of parameters or equation) by which aggregates such as "chemical and fertilizer mineral mining," "electrometallurgical products," and "primary nonferrous metals," can be disaggregated. An alternative approach is to expand the official input-output table by disaggregating the mineral cells into smaller cells. This permits varying the relationship between the sub-components and all of the cells outside a particular mineral or energy industry, but it imposes additional data collection and estimating requirements (distribution of purchases and sales) and "cements in" the base table relationships.

There is an implicit requirement, of course, that the physical amounts assumed to correspond with the dollar amount in the reference table actually do correspond. One cannot simply divide the dollar output of copper, say, by the quoted price of copper in the reference year. The actual amounts may be significantly different. Reports to the U.S. Census Bureau of the dollar value of shipments will not necessarily match those of the Bureau of Mines for the same year as to physical volume of production or shipments. At the same time, the physical production or shipment data reported to the Census Bureau, which should be consistent with the input-output table, may not be the most useful data to have as model output. The reference table's data specifications simply have to be investigated sufficiently to determine whether or not a model based on that table generates a valid index for the output series that is really desired.

A data requirement that is not essential to the use of existing tables but that is perceived by practitioners as critical to the utilization of input-output tables for projection purposes relates to the projection of changes in the "technical" (A-matrix) coefficients. Given the fact of continuing technological change, not to mention changes in product mix within input-output sectors, it is inevitable that these coefficients will change over time. A simplistic way of adapting them is merely to extrapolate the trend in the coefficients (possibly using an econometric approach) over successive input-output tables. This has severe drawbacks such as the lack of coefficient data over time, their dependence on stages of the business cycle, and the lack of suitable price deflators. An alternative way of projecting the technical coefficients is to bring to bear independent information on the relevant technological trends. Not only does this impose a massive information requirement, but it also necessitates the overcoming of the difficulty of translating a limited number of specific physical trends into aggregate monetary values that usually encompass more than those few interindustry trading relationships.

13 THE FUTURE

A broad view has been presented of the range of models and the methodologies applied to mineral and energy markets. This has been necessary because of the great diversity of models and model-related studies in this area. At present, a greater concentration of modeling effort appears to exist in analyzing specific sectors or special aspects of markets and industries rather than in modeling entire markets and systems. At the same time, the analysis of systems interaction has been evolving, particularly in the case of energy-environment-economy models. In this final chapter, an attempt is made to emphasize those modeling or quantitative areas which would appear to hold the greatest promise in future efforts.

SECTORAL PRICE MODELS

Continued uncertainty in mineral and energy investment and volatile commodity prices have maintained the need for price modeling and forecasting. Long-term price analysis is more concerned with shifting trends or structural breaks which seem to be related to major shifts in the world economy (Badillo, et al. 1997, Perron, 1989). The evaluation of these trends is important not only for evaluating mining investments, but also for assessing the impacts of mineral and energy product exports on developing countries. Among new applications are methods which endow these forecasts with risk-equivalent measures.

Medium-term price analysis is more concerned with price cycles arising either because of endogenous, disequilibrium adjustments in individual markets or because of exogenous shocks or adjustments related to national and international business cycles. The study of metal and material price cycles has continued since the seminal work of the NBER, and today has expanded using tests for cyclicality developed in the macroeconomic domain (Davutan and Roberts, 1994; Harvey, 1985; Labys et al., 1995; Labys and Kouassi, 1996; Moore, 1980). The prediction of these price cycles is important not only for efficient materials inventory management, but also for anticipating market movements. The relation of price cycles to macroeconomic adjustments has been concerned with inflationary price movements and the impacts of expansions and recessions on the mineral and energy industries (Bosworth and Lawrence, 1982; Fama and French, 1988; Chu and Morrison, 1984; Moore, 1988). Some of this research has been advanced by tests of duration and persistence and structural time series modeling.

Short-term price analysis is more related to hedging, speculation and trading on minerals and energy related futures and derivative markets. Research in this area has been stimulated by the time series and econometric methods that have been developed for analyzing financial market and price movements. Some of this work concerns random walks, chaos and fractile behavior (DeCoster et al., 1992; Frank and Stengos, 1989). Other works dwell on the possibility that ARFIMA models might better be applied when long memory is present in prices by using a fractional rather than an integer value for the degree of integration (Barkoulas et al., 1997, 1998); Cheung and Lai, 1994; Sowell, 1992). While this form of model emphasizes nonlinearity in price means, modeling based on nonlinearity in price variance has continued in the form of the variety of autoregressive hetereoscedastic models (Bollerslev, 1986; Engel, 1982; Kouassi et al. (1996), Subba Rao, 1981; Tong, 1983).

SECTORAL INVENTORY MODELS

Compared to time series models of price behavior, mineral and energy models occasionally have featured price relations linked to inventory adjustments. These have often proven effective because the relative price inelasticity of commodity demand and supply necessitate that inventory adjustments move markets towards equilibrium. While researchers have tended to reduce the price relations so as to eliminate the inventory variable, more recently attempts have been made to retain the inventory variable, if not to model inventory behavior directly. No doubt much of this work is the result of recently available inventory data.

Similar to the influence of macroeconomic analysis of price behavior, renewed interest in macroeconomic inventory behavior has spawned an interest in mineral and energy market inventories (Chikan, 1984; Chickan and Lovell, 1988; Eichenbaum, 1989). While forms of the accelerator theory received an initial interest in modeling inventories in aggregate, the production-cost-smoothing models consider that inventories are used to shift output to periods in which production costs are low, thus avoiding stockouts and reducing scheduling costs (Eckstein and Eichenbaum, 1985; Heo, 1996; Labys and Lord, 1992; Pindyck, 1994).

Inventory models which are more of a microeconomic nature have moved from a combined transactions-precautionary-speculative approach to that of nonlinear supply of storage, cost of carry, asset holding, and optimal competitive storage. Concepts of cost of carry are linked to the role of storage when speculation grows and backwardation is present. Here, inventory holding is said to lead to a convenience yield (Choe, 1992; Larson, 1994). The related optimal competitive storage model has been advanced by Williams and Wright (Williams and Wright, 1991; Wright and Williams, 1982, 1984). A number of attempts have been made to evaluate this model for potential applications in mineral and energy markets (Akayama and Trivedi, 1991; Gilbert, 1991; Kocagil, 1997; Thurman, 1988). Given the relative importance of inventories, not only for facilitating market adjustments, but also in affecting industry profitability at the materials purchasing level, commodity inventory modeling and forecasting is likely to improve mineral and industry commodity purchasing and management.

RISK ANALYSIS AND INVESTMENT MODELS

Above the forecasting of medium and long-run price behavior was emphasized as an important input into mineral and energy investment decision making. In such cases, price risk and uncertainty often is dealt with by forecasting price ranges rather than means or points. "What-if" kinds of model policy simulation analysis also can involve the "shocking" of certain variables with known event distributions and, for example, the observing of their resulting impact on prices or by performing some form of intervention analysis.

An alternative approach for assessing the impact of risk and uncertainty in a modeling context is to employ an expert systems framework. This often involves embedding a model in spreadsheet software and combining this model representation together with risk-analysis software using "expert systems," e.g. Harmon, et al., 1988; McDonald and Siegel, 1986; O'Leary, 1987; Sakarat, 1996; V-P Expert, 1993. Such methodologies typically employ discounted cash flow analysis including risk-adjusted discount rates. Capital asset or option pricing models are often employed based on the relative volatility of other assets. This can be coupled with probability analysis which reflects the likely impacts of different prices on different outcomes, i.e. see Torries (1998)

Applied to mineral or energy project evaluation, this approach can quantify market risk premiums through operational flexibility. In this sense, it allows the managerial decision process to identify the portion of project risk caused by operational flexibility. Different discount rates are evaluated until a final rate is found; this rate includes the incremental market risk which is then removed through operational flexibility using the mentioned expert systems. The operational flexibility often varies with production costs, where cash flows are cyclical at successive stages of project process. Investment decisions are thus based on an entry trigger value which must exceed the average total cost of the annualized full cost of making and operating the investment.

The future use of expert systems will depend on their effectiveness to accommodate both fluctuating prices and discount rates. At present, such systems can incorporate decision rules for mineral and energy investment and accordingly maximize net present value. It is also possible to identify and to capture other risk elements associated with mineral and energy projects.

MINERAL AND ENERGY MARKET MODELS

The most recent formal assessment of mineral and energy market models was concerned with larger scale models, many of which dealt with modeling global market interventions, i.e. see Labys and Lesourd (1991). Since then, modeling interests have become more specific, particularly in the application to market sector or component variables. Much of this has been due to the expansion of time series modeling. Attempts to incorporate such detailed sectoral representation into global models will probably await a new round of market crises. This has been the repeated experience of the past. At the present, major market modeling activities of interest involve (1) more rapid programming algorithms, and (2) greater attention to the noncompetitive structures of mineral and energy industries. Concerning more rapid programming algorithms, our interest is mainly with spatial price equilibrium models (Labys and Yang, 1997). Spatial equilibrium models can be linked to fixed-demand and fixed-supply forms of linear programming models via the Le Châtelier principle. As more and more constraints are added to the spatial system, the optimum solutions become less sensitive. In the case of disequilibrium spatial models, the linear programming transportation model can be shown to be the limiting case of either the Maxwell-Boltzmann or Bose-Einstein entropy algorithms. In addition, the linear programming transportation model can be made equivalent to the maximum flow problem, if the number of constraints on arcs are not numerous enough to yield a unique solution.

Recent advances in mathematical programming indicate that the evolution of spatial interaction models follows a path starting from linear programming and quadratic programming models. It is then followed by linear complementarity programming and its nonlinear versions. More recently, the most general spatial interaction model has evolved in the direction of variational inequalities. However, other advances involving market structure include spatial monopoly models and more recently the spatial Cournot model.

We should remember that the gravity formulation has been the backbone of spatial commodity models with applications to groups of commodities, transport networks, and demand distribution patterns. Such models can be derived by Wilson's entropy maximization or Erlanders' trip cost minimization or Smith's efficiency principle. Empirical applications of the gravity model can be implemented by the conditional logit model. Since all of these models are related to the linear programming model, they can all be linked to the spatial price equilibrium model. More than all other models, this approach has had success in analyzing policy decisions. What makes the Takayama-Judge (1971) model one of a few empirically relevant economic models is its computational advantage. It comes as no surprise that the Takayama-Judge model can serve as an essential and useful core for many interrelated spatial commodity models. The resulting unified model can be formulated as either a capacitated minimum cost or an entropy model.

Computational efficiency makes the spatial equilibrium models much easier to implement, whereas the computational advantage of the entropy or gravity models may not be as great. Lastly, spatial equilibrium models which are more market-oriented can be formulated to incorporate many important policy implications, e.g., spatial welfare and taxation effects. In contrast, the entropy model is not as applicable for policy implications since the economic interpretation of some parameters are less clear. The ultimate choice of a modeling method can be decided based on these properties.

In the case of the multi-commodity spatial equilibrium problem, the linear complementarity model may be preferred. However, in oligopolistic markets, the spatial equilibrium Cournot model may be chosen over its competitive model. In a planned economy, a linear programming model has the computational advantage. Furthermore, the entropy model can be applied to the random occurrence of commodity flows, while the maximum flow paradigm is adequate for rational maximum diffusion models, as is evidenced in traffic flow problems. Spatial logit models can also be formulated but with caution since they can have inconsistency problems. For a large-scale spatial equilibrium model, two-stage least-squares with principal components on the translog functions may be preferred to maintain internal consistency and to circumvent the negative degrees of freedom problem. For the moment, some of the largest and fastest models of this type, which also provide the most flexible programming model framework, can be solved using the variational inequality methodology.

Finally, there is the need to untangle more carefully the nature of mineral and energy market structures and the role played therein by different forms of market interventions such as regulated prices, subsidies, taxes and trade controls. The potential of multidisciplinary analysis in this area might be limited, unless the disciplines of quantitative political analysis and of industrial organization advance in this direction. Attention to market structure in mineral and energy models has been expanded by Pindyck (1978a,b), Salant et al. (1981), Kolstad (1982), Lord (1991) and others. We have thus seen the embodied structure of the petroleum market advance from simple monopoly to rather complex forms of oligopoly. Greater attention to mineral and energy market structure appears in Labys (1980a) and Lord (1991). A lesser amount of work has taken place regarding the role of market interventions and disruptions. Mineral and energy models are thus likely to improve in the direction of offering a more realistic picture of these two related aspects of market structure.

ENERGY-ENVIRONMENT-ECONOMY MODELS

A growing model direction mentioned earlier involves measuring the economic impacts of mineral and energy commodities on the macroeconomy and the resulting environmental interactions. This approach would demonstrate unfavorable effects likely to occur over the long term, so as to provide incentives for policy makers to react responsibly. This modeling area can be considered diverse and includes many different kinds of interactions: (1) impacts of energy production and consumption on the environment and the resulting pollution and pollution-abatement costs; (2) interactions between production technologies, the environment and the global economy; (3) relation of climate systems to agricultural production and the economy; (4) greenhouse effect models including carbon taxation and economic impacts; (5) the influence of environmental regulations on mineral and energy international trade patterns; and (6) impacts of increasing emission standards on regional structural adjustments in the minerals, coal and other energy industries.

A number of modeling studies and reviews have taken place which attempt to define this modeling area and to address the many modeling

possibilities (Beaver and Huntington, 1992; Barker et al., 1993; . Dowlatabedi and Morgan, 1993; Farzin, 1996; Hope, 1993; Hourcade, 1993; Manne and Richels, 1992; Peck and Tiesberg, 1993; Wilson and Swisher, 1993). A recently significant research effort has examined how carbon tax policies could reduce CO_2 emissions and global warming (Hoel, 1993, 1994; Farzin, 1995; Farzin and Tahvonen, 1996; Wirl, 1995). While most such s
tudies have concentrated on the energy industry, several also deal with different facets of the minerals industry (Eggert, 1994; Ruth, 1995a,b; Weston and Ruth, 1997; Williams et al., 1987; Yoshiki-Gravelsins, et al. 1993).

A major problem to be confronted with these modeling efforts are the uncertainties concerning not only long-term structural relationships but also the fundamentals of the atmospheric transfer process. Accordingly, environmental impacts are not easily quantifiable, and the economic or societal responses are unclear. While much is known about the long-term costs of abatement, less is known about technology development. Future modeling efforts have thus to deal collectively with the nature of environmental impacts, pollution abatement policies and economic costs, new technologies and the uncertainties of the economic, energy and climatic systems.

Also of increasing importance are modeling efforts which concentrate particularly on the international trade aspects of environmental and economic interactions (Klein, 1992; Perroni and Wigle, 1994; Puttock and Sabourin, 1992; Sutton, 1988; Tobey, 1990). These models recognize that environmental regulations can be applied at the international trade level to improve the quality of the mix of goods traded. In some cases, though, the regulations have been used as hidden forms of trade barriers. Because the World Trade Organization has no direct mechanism for dealing with environmental trade disputes, the possibilities for modeling the impacts of particular tariff and non-tariff barriers are complex. However, such modeling efforts could have the benefit of demonstrating how global trade expansion might take place in the context of improved environmental standards.

POSTSCRIPT

Finally, it goes without saying that in times of mineral and energy market crises, greater attention becomes directed towards market modeling and particularly modeling market risk and uncertainty. Advances in the professional field of mineral and energy modeling should not depend on such crises. The models and findings which have come before us should be used to provide new advancements so that this field of economic modeling will continue to improve and to grow.

REFERENCES

Aashtiani, M.Z., and T.L. Magnanti. 1981. Equilibria on a Congested Transportation Network, *SIAM Journal on Algebraic and Discrete Methods,* 2: 213-26.

Adams, F.G. 1972. The Impact of the Cobalt Production from the Ocean Floor: Working Paper, Wharton School of Finance, Philadelphia, PA.

Adams, F.G., and J.M. Griffin. 1972. An Econometric Linear Programming Model of the U.S. Petroleum Industry• *Journal of the American Statistical Association,* 67: 542-551.

Adams, F.G. and J.R. Behrman. 1982. *Commodity Exports and Economic Development.* Lexington, MA: Heath Lexington Books.

Adams, F.G., E.A. Kroch, and V. Didzivlis. 1991. The Linkages between Markets for Petroleum Products and the Market for Crude Oil.• In O. W.C. Labys and J.B. Lesourd (eds), *International Commodity Market Models, London: Chapman & Hall.*

Adelman, M.A. 1983. *Energy Resources in the Uncertain Future: Oil Supply Forecasting.* Cambridge, MA: Ballinger Publishing Co.

AES Corporation. 1990. *An Overview of the Fossil 2 Model,* U.S. Department of Energy, Office of Policy and Evaluation, Arlington, VA.

Akiyama, T. and P.K. Trivedi. 1991. Stabilizing Imported Food Prices for Small Developing Countries: Any Role for Futures? Discussion Paper, International Economics Department, World Bank, Washington, D.C.

Almon, C., Jr. M.B. Buckler, L. M. Horowitz and T.C. Reimbold. 1974. *Interindustry Forecasts of the American Economy* Lexington, MA: Heath Lexington Books.

Amit, R. and M. Avriel. 1982 *Perspective on Resource Commodity Modeling: Minerals and Energy.* Cambridge, MA: Ballinger Publishing Co.

Anderson, D. 1972. Models for Determining Least-Cost Investments in Electricity Supply, *Bell Journal of Economics and Management Science,* 3: 267-99.

Aperjis. D.G. 1981. *Oil Markets in the 1980's.* Cambridge, MA:Ballinger Publishing Co.

Archibald, R. and R. Gillingham. 1980. An Analysis of the Short-run Consumer Demand for Gasoline Using Household Survey Data. *Review of Economics and Statistics,* 62:622-8.

Armington, P.S. 1969. A Theory of Demand for Products Distinguished by Place of Production, *IMF Staff Papers,*16:159-76

Arrow, K.J. and S. Chang. 1982. Optimal Pricing, Use and Exploration of Uncertain Nature Resource Stocks, *Journal of Environmental Economics and Management,* 9:1-10.

Assimakopoulos, V. 1992. Residential Energy Demand Modelling in Developing Regions: The Use of Multivariate Statistical Techniques, *Energy Economics*, 14:57-63.

Ayres, R. U., 1969. *Technological Forecasting and Long-Range Planning*, New York: McGraw-Hill.

Badillo, D., Labys, W.C. and Yang-ru Wu. 1997. Identifying Trends and Breaks in Commodity Prices, *European Journal of Finance, forthcoming*.

Balestra, P. and M. Nerlove. 1966. Pooling Cross-Section and Time Series Data in the Estimation of a Dynamic Model: The Demand for Natural Gas, *Econometrica*, 34:585-612.

Baltagi, B.H. and J.M. Griffin. 1983. Gasoline Demand in the OECD: An Application of Pooling and Testing Procedures. *European Economic Review*, 22:117-137.

Baltagi, B.H. and J.M. Griffin. 1988. A General Index of Technical Change. *Journal of Political Economy*, 96:20-41.

Ban, K. 1996. Energy Conservation and Economic Performance in Japan: An Econometric Approach, *Models for Energy Policy*, In J.B. Lesourd, J. Percebois, and F. Valette (eds), New York: Routledge.

Barker, T., Bayles, S., and P. Madgen. 1993. A UK Carbon/Energy Tax: the Macroeconomic Effects, *Energy Policy*, 21: 296-308.

Barkoulas, J., W.C. Labys and J. Onochie. 1997. Fractional Dynamics in International Commodity Prices. *Journal of Futures Markets, 17:161-189*

Barkoulas, J., W.C. Labys, and J. Onochie. 1998. Long Term Memory in Commodity Futures Prices, *The Financial Review*, forthcoming.

Barnett, H.J., 1950. *Energy Uses and Supplies, 1939, 1947, 1965*. IC 7582, U.S. Bureau of Mines, U.S. Department of Interior, Washington, DC.

Barrett, S.A. 1982. Modeling Coal Supply: An Econometric Analysis of Competing Mining Methods. Unpublished manuscript, Electric Power Research Institute, Palo Alta.

Basile, P.S., 1979. The IIASA Set of Energy Models, Working Paper, International Institute for Applied Systems Analysis, Laxenburg.

Batten, D.F., and B. Johansson. 1985. Price Adjustments and Multiregional Rigidities in the Analysis of World Trade, *Papers of the Regional Science Association*, 56: 145-66.

Batten, D.F., and L. Westin. 1989. Modeling Commodity Flows on Trade Networks: Retrospect and Prospect. Department of Economics Report No. 194, University of Umea.

Baughman, M. L. and P.L. Joskow. 1974. A Regionalized Electricity Model, MIT-EL 75-005, MIT Energy Lab, Cambridge, MA.

Beaver, R.D., 1992. *Technical Summaries of EMF 12 Models*, Energy Modeling Forum, Stanford University, Stanford, CA.

Beaver, R.D., and G. Huntington. 1992. A Comparison of Aggregate Energy Demand Models for Global Warming Policy Analyses, *Energy Policy*, 20: 568-574.

Beaver,R. 1993. Structural Comparison of the Models in EMF12. *Energy Policy*, 21:858-867

Beck, R., and J.L. Solow. 1994. Forecasting Nuclear Power Supply with Bayesian Autoregression, *Energy Economics*, 16: 185-192.

Beckman, M.J., and T.F. Golob. 1972. A Critique of Entropy and Gravity in Travel. In *Traffic Flow and Transportation*, G.F. Newell (ed) New York: American Elsevier.

Beenstock, M. and P. Willcocks. 1983. Energy Consumption and Economic Activity in Industrialized Countries, *Energy Economics*, 3: 225-32.

Beltzer, D., and C. Almon. 1972. Forecasts of U.S. Petroleum Demand: An Interindustry Analysis, Research Report 4, Bureau of Business and Economic Research, University of Maryland.

Bentzen, J. and T. Engsted. 1993a. Expectations, Adjustment Costs, and Energy Demand, *Resource and Energy Economics,* 15: 371-85.

Bentzen, J. and T. Engsted. 1993b. Short and Long Run Elasticities in Energy Demand: A Cointegration Approach, *Energy Economics*, 15: 9-16.

Bentzen, J. 1994. An Empirical Analysis of Gasoline Demand in Denmark using Cointegration Techniques, *Energy Economics*, 16:139-144.

Berndt, E.R. 1983. Quality Adjustment, Hedonics, and Modern Empirical Demand Analysis, *Price Level Measurement*. Proceedings from a Conference sponsored by Statistics Canada. W.E. Diewert and C. Montmarquette, (eds). Minister of Supply and Services, Ottawa, Canada. pp 817-863.

Berndt, E.R. and B. Field (eds). 1988. *Measuring and Modeling Natural Resource Substitution.* Cambridge, MA: MIT Press.

Berndt, E. R. and D. O. Wood. 1975. Technology, Prices and the Derived Demand for Energy, *Review of Economics and Statistics*, 111: 259-68.

Berndt, E.R., and D.O. Wood. 1979. Engineering and Econometric Interpretations of Energy-Capital Complementarity,• *American Economic Review*, 69: 342-354.

Berndt, E., M. Fuss, and L. Waverman. 1980. Dynamic Adjustment Models of Industrial Energy Demand: Empirical Analysis for U.S. Manufacturing, 1947-74. Research Project No. 683-1, Electric Power Research Institute, Palo Alto.

Berndt, E.R. and D.O. Wood. 1982. The Specification and Measurement of Technical Change in U.S. Manufacturing, *Advances in the Economics of Energy and Resources.* Vol. 4, J. R. Mooney, (ed)., Greenwich, CT: JAI Press Inc., 199-221.

Berndt, E.R. and D.O. Wood. 1984. Energy Price Changes and the Induced Revalution of Durable Capital in U.S. Manufacturing During the OPEC Decade, MIT Energy Lab, Cambridge, MA. 84-003

Bernkopf, R.L. 1985. *Domestic Coal Distribution: An Interregional Programming Model for the U.S. Coal Industry.* Greenwich, CN: JAI Press, Inc.

Bing, P.C. 1987. *A Model of Exploratory Drilling in Western Canada,* IAEE Ninth International Conference, Calgary, 6-8 July, pp. 707-17.

Birol, F. and N. Guerer. 1993. Modeling the Transport Sector Fuel Demand for Developing Countries. *Energy Policy,* 21: 1163-1173.

Bjerkholt, O., O. Olsen, and J. Vislie. 1990. *Recent Modeling Approaches Applied in Energy Economics.* London: Chapman & Hall.

Blinder, A.S. and S.J. Maccini. 1991. Taking Stock: A Critical Assessment of Recent Research on Inventories, *Journal of Economic Perspectives,* 5:73-96.

Blitzer, C., A. Meeraus, and A. Stoutjesdijk. 1975. A Dynamic Model of OPEC Trade and Production, *Journal of Development Economics,* 2: 319-335.

Blitzer, C.B. and R.S. Eckaus. 1985. Projecting Energy Demand in Developing Countries: A General Equilibrium Approach. Energy Lab Report EL85-010WP, MIT Energy Lab, Cambridge, MA.

Bocoum, B. and W.C. Labys. 1993. Modeling the Economic Impacts of Further Mineral Processing, *Resources Policy,* 19:1-27.

Bohi, D.R. 1981. *Analyzing Demand Behavior: A Study of Energy Elasticities,* Baltimore/London: The Johns Hopkins University Press for Resources for the Future.

Bohi, D.R. and M.A.Toman.1986. Analyzing Nonrenewable Resource Supply Research 1220-1, Electric Power Research Institute, Palo Alto, CA.

Bohi, D. 1989. *Energy Price Shocks and Macroeconomic Performance.* Washington, DC: Resources for the Future.

Bollerslev, T. 1986. Generalized Autogregressive Conditional Heteroscedasticity. *Journal of Econometrics,* 31:307-327.

Bollerslev, T. 1988. On the Correlation Structure for the Generalized Autoregressive Conditional Heteroscedastic Process, *Journal of Time Series Analysis,* 9:121-131.

Borges, A.M. 1986. Applied General Equilibrium Models: An Assessment of their Usefulness. *OECD Economic Studies*, 7: 7-43.

Borges, A.M., and G. Lawrence. 1984. Decomposing the Impact of Higher Energy Prices. In H.E. Scarf and W. B. Shoneu (eds) *Applied General Equilibrium Analysis*. New York: Cambridge University Press.

Bosworth, B. and R.Z. Lawrence. 1992. *Commodity Prices and the New Inflation*, Washington D.C.: The Brookings Institute.

Booth, G.G., Kaen, F.R., and Koveos, P.E. 1982. Persistent Dependence in Gold Prices, *Journal of Financial Research*, 5:85-93.

Box, G. and G. Jenkins. 1970. *Time Series Analysis*. San Francisco: Holden Day.

Boyd, G.A. 1992. Forecasting Industrial Energy Use. In T. Sterner (ed) *International Energy Economics*. London: Chapman & Hall.

Bresnahan, T.F. an dV.Y. Suslow. 1985. Inventories as an Asset: The Volatility of Copper Prices, *International Economic Review*, 26:409-24.

Broadman, H.G. 1985. Incentives and Constraints on Exploratory Drilling for Petroleum in Developing Countries, *Annual Review Energy*, 10: 217-49.

Brocker, J. 1984. How do international trade barriers affect interregional trade? In A.E. Andersson, W. Isard, and T. Puu. (eds) *Regional and industrial development: theories, models, and empirical evidence*, Amsterdam: North-Holland.

Brocker, J. 1988. Interregional Trade and Economic Integration: A Partial Equilibrium Analysis, *Regional Science and Urban Economics*, 18: 261-81.

Brocker, J. 1989. Partial Equilibrium Theory of Interregional Trade and the Gravity Model. *Papers of the Regional Science Association*, 66: 7-18.

Brooke, A., D. Kendrick, and A. Meerus. 1988. *GAMS*, San Francisco: Scientific Press.

Brown, M., A. Dammert, A. Meeraus, and S. Stoutjesdijk. 1983. *Worldwide Investment Analysis: The Case of Aluminum*. Staff Working Paper 603, World Bank, Washington, D.C.

Burbridge, J., and A. Harrison. 1984. Testing for the Effects of Oil Price Rises using Vector Autoregressions, *International Economic Review*, 459-484.

Burrows, J. 1972. *Analysis and Model Simulations of the Non-Ferrous Metal Markets; Aluminum*. Special Report, Charles River Associates, Cambridge MA.

Burrows, J. 1971. *Cobalt: An Industry Analysis*. Lexington, Mass.: Heath Lexington Books.

Cairns, R.D. 1986. *The Economics of Energy and Mineral Exploration: A Survey.* IIASA, Laxenburg, Austria.

Camos, M. 1986. *MEDEE 3, Modele de Demande de Energie pour Europe*, Paris: Lavoisier, Collection Tec-Doc.

Campbell, T.C., M.J. Hwang, and F. Shahrokh. 1980. Spatial Equilibrium in the U.S. Coal Industry, *Energy Economics*, 2: 230-236.

Capros, P., Karadeloglou,P. Mantzos, L. and G. Mentzas. 1996. The Energy Model MIDAS. *In Models for Energy Policy*, J.B. Lesourd, J. Percebois, and F. Valette, (eds) *New York*: Routledge.

Capros, P., Karadeloglou, P., and G. Mentzas. 1990. New Developments for the MIDAS Medium-term Energy Modeling Project of the EEC: the Energy Supply Model and the Supply-Demand-Pricing Linkage, Paper presented at the 12th Triennial Congress on Operations Research IFORS 90, Athens 25-29 June 1990.

Capros, P., Karadeloglou, P. and G. Mentzas. 1992. Energy Policies in a Macroeconomic Model. In T. Sterner (ed) *International Energy Economics*, London: Chapman & Hall.

Carson, J., W. Christian, and G. Ward. 1981. The MIT World Oil Model, Energy Lab Report, WP81-027, MIT Energy Lab, Cambridge, MA.

Carter, A.P. and A. Brody. 1970. *Contributions to Input-Output Analysis.* North-Holland: Amsterdam.

Cazalet, E.J. 1975. *SRI-Gulf Energy Model: Overview of Methodology.* Stanford Research Institute, Menlo Park, CA.

Charles River Associates. 1978. *The Economics and Geology of Mineral Supply: An Integrated Framework for Long-Run Policy Analysis.* Report N. 327, Charles River Associates, Boston.

Charles River Associates. 1976. *Modeling Analysis of Supply Restrictions in the Minerals Industry.* Charles River Associates, Boston.

Charles River Associates. 1976. *Econometric Model of the World Nickel Industry.* Charles River Associates, Boston

Charles River Associates 1982. *CRA/EPRI Coal Market Analysis System.* EPRI Report EA-907 (in 4 vols.) Electric Power Research Institute, Palo Alto, CA.

Cheung, Y.W., and Lai, K.S. 1993. Do Gold Market Returns Have Long Memory? *Financial Review*, 128:181-202.

Chichilnisky, G., 1981. Oil Prices, Industrial Prices and Outputs: A General Equilibrium Macro Analysis. Working Paper, United Nations Institute for Training and Research, New York.

Chikan, A. 1984. *New Results in Inventory Research.* Amsterdam: Elsevier-Science.

Chikan, A. and M. Lovell, eds. 1988. *The Economics of Inventory Management.* Amsterdam: Elsevier-Science.

Choe, B.J. 1992. The Precautionary Demand for Commodity Stocks, WPS 935, International Economics Department, World Bank, Washington, D.C.

Choksi, A.M., A. Meeraus, and A.J. Stoutjesdijk. 1983. *The Planning of Investment Programs in the Fertilizer Industry.* Baltimore: Johns Hopkins University Press.

Choukroun, J.M. 1975. A General Framework for the Development of Gravity-type Trip Distribution Models, *Regional Science and Urban Economics,* 5: 177-202.

Christensen, L.R. and W.H. Greene. 1976. Economies of Scale in U.S. Electric Power Generation, *Journal of Political Economy,* 84:655-676.

Chu, K.Y. and T.K. Morrison. 1984. The 1981-82 Recession and Non-Oil Primary Commodity Prices, *IMF Staff Papers,* 31:93-140.

Church, A.M. 1981. *Taxation of Nonrenewable Resources.* Lexington, MA: Heath Lexington Books.

Clark, J.P, and A. Church. 1981. *Process Analysis Modeling of the Stainless Steel Industry.* Board on Mineral and Energy Resources, Committee on Nonfuel Mineral Demand Relationships, National Academy of Sciences, Washington, DC.

Clark J.P., and S.C. Mathur. 1981. *An Econometric Analysis of the Substititution Between Copper and Aluminum in the Electric Conductor Industry.* Board on Mineral and Energy Resources. Committee on Nonfuel Mineral Demand Relationships. National Academy of Sciences, Washington, DC.

Clark, J.P., J. Tribendis, and J. Elliott. 1981. An Analysis of the Effects of Technology, Policy and Economic Variables on the Ferrous Scrap Market in the United States, Working Paper, Department of Materials Science and Engineering, Massachusetts Institute of Technology.

Clark, P., Coene, P. and D. Logan. 1981. A Comparison of Ten U.S. Oil and Gas Supply Models, *Resources and Energy,* 3: 297-335.

Cohen, B.C. and W.C. Labys. 1994. CO_2 Emissions and Concentration Coefficients Revisited. *International Journal of Global Energy Issues,* 6:366-370.

Cohen, B.C. and W.C. Labys. 1996. Uncertainty, Carbon Dioxide Concentrations and Fossil Fuel Use, *International Journal of Environment and Pollution,* Vol. 6, No.1.

Cohen, B.C. and W.C. Labys. 1997. An Econometric Approach to CO_2 Concentration Modeling. In F. Carlevaro and J.B. Lesourd, (eds.) *Measuring and Accounting Environmental Nuisances and Benefits:Essays in Environmetrics,* London: Macmillan.

Commoner, B. 1983. *The Politics of Energy.* New York: Knopf.

Consadine, T.J. 1991. Technical Change, Relative Prices and International Substitution, In W.C. Labys and J.B. Lesourd (eds.). *International Commodity Market Models,* London: Chapman & Hall.

Copithorne, L. W. 1973. *The Use of Linear Programming in the Economic Analysis of a Metal Industry: The Case of Nickel.* Department of Economics Monograph, University of Manitoba, Winnipeg, Canada.

Cottle, R.W., and G.B. Dantzig. 1968. Complimentary Pivot Theory of Mathematical Programming, *Linear Algebra and Its Applications,* 1: 103-25.

Cox, J.C. and A.W. Wright. 1976. The Determinants of Investments in Petroleum Reserves and their Implications for Public Policy, *American Economic Review,* 66:153-67.

Cremer, J. and M. Weitzman, 1976. OPEC and the Monopoly Price of World Oil, *European Economic Review,* 8: 155-64.

Criqui, P. 1996. Methodological Advances in Energy Modeling: 1970-90" In J.B. Lesourd, J. Percebois, and F. Valette (eds.), *Models for Energy Policy,* New York: Routledge.

Criqui, P. 1996. International markets and energy prices: the POLES model, In J.B. Lesourd, J. Percebois, and F. Valette (eds.) ,*Models for Energy Policy,* New York: Routledge.

Crissey, B.L. 1975. A Rational Framework for the Use of Computer Simulation Models in a Policy Context, Ph. D. Dissertation. Johns Hopkins University, Baltimore, MD.

Cromwell, J., Labys, W.C. and M. Terraza. 1994a. *Univariate Tests for Time Series Models.* Thousand Oaks: SAGE.

Cromwell, J., Hannon, M., Labys, W.C. and M. Terraza. 1994b. *Multivariate Tests for Time Series Models.* Thousand Oaks: SAGE.

Crowson, P.C.F. 1981. Demand Forecasting: A Mining Company's Approach, Board on Mineral and Energy Resources, Committee on Nonfuel Minerals Demand Relationships, National Academy of Sciences, Washington, DC.

Cutler, J.S., and B. Schachter. 1989. The Investment Decision: "Estimation of Risk and Risk Adjusted Discount Rate. *Financial Management,* 18: 13-22.

Dahl, C.A. 1993. A Survey of Energy Demand Elasticities in Support of the Development of NEMS. Memorandum 98, Gotenborg University, Sweden.

Dahl, C.A. 1986. Gasoline Demand Survey, *Energy Journal,* 7:67-82.

Dahl, C.A., and T. Sterner. 1990. The Pricing of and the Demand for Gasoline: A Survey of Models, Memorandum 132, Götenborg University, Sweden.

Dahl, C.A. and T. Sterner. 1991. Analyzing Gasoline Demand Elasticities, *Energy Economics*, 13:127-139.

Dalcantara, G. and A. Italianer. 1982. European Project for a Multinational Macro-sectoral Model, MS11, XII/759/82, Commission of the European Communities, Brussels.

Dammert, A. 1980. Planning Investments in the Copper Sector in Latin America. In W. Labys, M. Nadiri, and J. Nunez del Arco (eds.) *Commodity Markets and Latin American development: A Modeling Approach*, New York: National Bureau of Economic Research.

Dammert, A., and J. Chrabra. 1987. *Long-term Prospects for the Lead and Zinc Industries*. EPDCS International Working Paper, World Bank, Washington, D.C.

Dammert, A., and S. Palaniappan. 1985. *Modeling Investments in the Copper Sector*. Austin, TX: University of Texas Press.

Dantzig, G.B. 1949. Programming in Linear Structure, *Econometrica*, 17: 73-74.

Dantzig, G.B. 1963. *Linear Programming and Extensions*. Princeton, NJ: Princeton University Press.

Dargay, J.M. 1988. *The Demand for Petroleum Products in Europe*, Working Paper, Oxford Institute of Energy Studies, Oxford.

Dartmouth Systems Dynamics Group. 1977. FOSSIL1: Introduction to the Model, Working Paper, Dartmouth College, Hanover, NH.

Davidoff, R.L. 1980. Supply Analysis Model (SAM): A Minerals Availability System Methodology, IC 8820, US Bureau of Mines, Washington, DC.

Davidson, R., W.C. Labys and J.B. Lesourd. 1998. Wavelet Analysis of Commodity Price Behavior, *Computational Economics, 11:103-128*.

Davutyan, N. and M. Roberts. 1994. Cyclicality in Resource Prices. *Resources Policy*, 20:49-57.

Day, R.H. 1973. Recursive programming models: A brief introduction, In G. Judge and T. Takayama *(eds.) Studies in Economic Planning Over Space and Time*, Amsterdam: North-Holland.

Day, R.H. and W.K. Tabb. 1972. A Dynamic Microeconomic Model of the U.S. Coal Mining Industry, SSRI Research Paper. University of Wisconsin, Madison.

Day, R.H. and J.P. Nelson. 1973. A Class of Dynamic Models for Describing and Projecting Industrial Development, *Journal of Econometrics*, 2: 155-190.

Day, R.H. and E. Sparling. 1977. Optimization Models in Agricultural and Resource Economics, In G.G. Judge, ed, *Quantitative Models in Agricultural Economics*. Vol. 2 in the AAEA Survey of Agricultural Economics Literature, University of Minnesota, Minneapolis.

Deam, R.J., J.G. Hale, J.R. Isaac, J. Leather, F.M. Carroll, P.V. Slee, P.C. Ward, and P.L. Watson. 1974. World Energy Modeling: Concepts and Methods and Preliminary Results from the Petroleum/natural gas Model. In *Energy Modeling*, U.S. National Science Foundation and the Energy Research Unit, Queen Mary College, London: IPC Science and Technology Press.

DeCoster, G.P., W.C. Labys and D.W. Mitchell. 1992. Evidence of Chaos in Commodity Futures Prices, *Journal of Futures Markets*, 12:291-305.

DeGenring, P.C., and J.R. Jackson. 1980. An Econometric-Engineering Analysis of the Impact of the National Energy Act on U.S. Commercial Sector Energy Demand, *Energy Modeling Studies and Conservation*. Economic Commission for Europe, New York: Pergamon Press for the United Nations,303-318.

Deimezis, N. 1996. ERASME: A Short-term Energy Forecasting Model for the European Community, In J.B. Lesourd, J. Percebois, and F. Valette, (eds.) *Models for Energy Policy*, New York: Routledge.

De Mesnard, L. 1996. Biproportional Methods and Interindustrial Dynamics: Application to Energy Demand in France, In J.B. Lesourd, J. Percebois, and F. Valette (eds.), *Models for Energy Policy*, New York: Routledge.

Deonigi, D. 1977. A World Trade Model, In *Proceedings of the Workshop on World Oil Supply-Demand Analysis*. Brookhaven National Laboratory, New York.

Desai, M. 1966. An Econometric Model of the World Tin Economy, *Econometrica*, 34: 105-134.

Desbarats, C.M. 1989. Empirical Modeling of Canadian Petroleum Exploration Activity, OIES Papers on Energy Economics, Oxford Institute for Energy Studies, Oxford.

DeSouza, G.R. 1981. *Energy Policy and Forestry*. Lexington: Heath Lexington Books.

Detemmerman, V., E. Donni and P. Zagamé. 1996. Energy Tax Increases as a Way to Reduce CO_2 Emissions (HERMES) In J.B. Lesourd, J. Percebois, and F. Valette (eds.) *Models for Energy Policy*, New York: Routledge.

Devanney III, J.W., and M.B. Kennedy. 1980. A Short Run Model of the World Petroleum Network Based on Decomposition. In W.T. Ziemba, S.L. Schwartz, and E. Koenigsberg (eds.). *Energy Policy Modeling*, Boston: Martinus Nijhoff.

Devarajan, S. and A.C. Fisher. 1982. Exploration and Scarcity. *Journal of Political Economy*, 90:6.

Diewert, W.E. and T.J. Wales. 1987. Flexible Functional Forms and Global Curvature Conditions, *Econometrica*, 55:43-68.

Donnelly, W.A. 1987. *The Econometrics of Energy Demand: A Survey of Applications*, New York: Praeger.

Dorfman, R.,P. Samuelson, and R. Solow. 1950. *Linear Programming and Economic Analysis*, New York: McGraw-Hill Book Company.

Dowlatabadi, H. and M.G. Morgan. 1993. A Model Framework for Integrated Studies of the Climate Problem, *Energy Policy*, 21:209-221.

Driehuis, W. (ed) 1976. *Primary Commodity Prices: Analysis and Forecasting*. Rotterdam: Rotterdam University Press.

Drollas, L.P. 1984. The Demand for Gasoline: Further Evidence, *Energy Economics*, 6: 71-82.

Dubin, A. and D.L. McFadden. 1984. An Econometric Analysis of Residential Electric Appliance Holdings and Consumption, *Econometrica*, 52: 345-62.

DuPree, W.G. Jr., and J. West. 1972. *United States Energy Through the Year 2000*, U.S. Bureau of Mines, Dept. of Interior, Washington, DC.

Duncan, R.C. 1984. The Outlook for Primary Commodities, 1984 to 1995, Staff Working Paper No. 11, World Bank, Washington, DC.

Dutton, C. M. 1982. *Modeling the International Steam Coal Trade*. Energy Research Group Paper EDP21, Cavendish Laboratory Cambridge University, Cambridge.

Eckbo, P.L., Jacoby, H,D. and J.L. Smith. 1978. Oil Supply Forecasting a Disaggregated Process Approach, *Bell Journal of Economics*, 9: 218-238.

Eckstein, A. and M.S. Eichenbaum. 1985. Inventories and Quality Constrained Equilibrium: The U.S. Petroleum Industry, 1947-72, In T. Sargent (ed.) *Energy Foresight and Strategies*. Washington, D.C.:Resources for the Future.

Economic Commission for Europe (ECE), 1980. *Energy Modeling Studies and Conservation*. New York: Pergamon Press for the United Nations.

Eden, R.J. 1993. World Energy to 2050: Outline Scenarios for Energy and Electricity, *Energy Policy*, 21: 231-237.

Eggert, R.G. 1994. Mining and the Environment: An Introduction and Overview, In R.G. Eggert (ed.) *Mining and the Environment: Interntational Perspectives on Public Policy*, Washington, D.C.: Resources for the Future, 1-20.

Eichenbaum, M.S. 1989. Some Empirical Evidence on the Production Level and Production Cost Smoothing Models of Inventory Investment, *American Economic Review*, 79:853-64.

Elliott, J.F., and J.P. Clark. 1977. Mathematical Modeling of Raw Materials and Energy Needs of the Iron and Steel Industry in the USA. Open-file Report 32-78. U.S. Bureau of Mines, Washington, DC.

Energy Modeling Forum (EMF) 1978. *Coal in Transition: 1980-2000*, Vols. 1 and 2, Energy Modeling Forum, Stanford University, Stanford, CA.

Energy Information Administration (EIA), 1979. *Documentation of the Project Independence Evaluation System*, Vols. 1-4. U.S. Department of Energy. Washington, D.C.: U.S. Government Printing Office.

Engle, R. 1982. Autoregressive Conditional Heteroscedasticity with Estimates of the Variance of the U.K. Inflation, *Econometrica*, 50:987-1007.

Engle, R.F. and T. Bollerslev. 1986. Modeling the Persistence of Conditional Variances, *Econometric Reviews*, 5,1:1-50.

Enke, S. 1951. Equilibrium Among Spatially Separated Markets: Solution by Electric Analogue, *Econometrica*, 19: 40-48.

Epple, D.N. 1975. *Petroleum Discoveries and Government Policy: an Econometric Study of Supply*. Cambridge, MA: Ballinger Publishing Co.

Epple, D.N. 1985. The Econometrics of Exhaustible Resource Supply: *A Theory and Application, In* Thomas J. Sargent *(eds.) Energy, Foresight and Strategy*. Resources for the Future, Washington, DC.

Erikson, E. and R.W. Spann. 1971. Supply Response in a Regulated Industry—Case of Natural Gas. *Bell Journal of Economic, 2:116-128.*

Erlander, D. 1977. Accessibility, Entropy, and the Distribution and Assignment of Traffic, *Transportation Research*, 11: 149-53.

Exxon Corporation, 1977. World Energy Outlook, Exxon Public Affairs Department, New York.

Fabian, R. 1963. Process Analysis of the U.S. Iron and Steel Industry, In A.S. Manne and H.M. Markowitz (eds.) *Studies in Process Analysis: Economy-Wide Production Capabilities*, Cowles Foundation for Research in Economics Monograph 18, New York: John Wiley & Sons.

Fama, E.F. and K.R. French. 1988. Business Cycles and the Behavior of Metals Prices, *Journal of Finance*, 43:1075-1088.

Fanyu, Pei. 1996. Econometric Estimates of Metal Demand Elasticities, M.S. Thesis, Colorado School of Mines, Boulder.

Farzin, Y.H. 1995. Optimal Timing of Global Warming and its Control Policy, Fodozione Eni Enrico Mattei (FEEM), Working Paper 86.95, Milan.

Farzin, Y.H. 1996. Optimal Pricing of Environmental and Natural Resource Use With Stock Externalities, *Journal of Public Economics*, 62:31-57.

Farzin, Y.H. and O. Tahvonen. 1996. Global Carbon Cycle and the Optimal Time Path of a Carbon Tax, *Oxford Economic Papers*, 48:215-228.

Federal Energy Administration. 1974. *Project Independence Report*, Washington, DC: U.S. Government Printing Office.

Fericelli, J. and J.B. Lesourd (eds.) 1985. *Energie: Modelisation et Econometrie.* Paris: Economica.

Fettweis, G.B. 1983. Considerations on Coal Deposits as Basis of Coal Production, International Congress of Carboniferous Stratigraphy and Geology, Madrid, Spain, 93-110.

Fisher, F. and C. Kaysen. 1962. *A Study in Econometrics: The Demand for Electricity in the United States*, Amsterdam: North-Holland.

Fisher, F. 1964. *Supply and Costs in the U.S. Petroleum Industry*, Washington, D.C.: Resources for the Future.

Fisher, F.M., P.H. Cootner, and M. Bailey. 1972. An Econometric Model of the World Copper Industry, *Bell Journal of Economics,* 3: 568-609.

Foley, P., and J. Clark. 1981. US Copper Supply - An Economic/engineering Analysis of Cost-supply Relationships, *Resources Policy*, 7:171-187.

Forrester, J. 1971. *World Dynamics.* Cambridge, MA: Wright-Allen Press.

Fox, K.A. 1953. A Spatial Equilibrium Model of the Livestock Feed Economy, *Econometrica,* 21: 546-66.

Fox, K.A. 1963. Spatial price equilibrium and process analysis in the food and agricultural sector. In A. Manne and H. Markowitz.(eds) *Studies in Process Analysis*, New York: John Wiley & Sons.

Fox, K. A., and R. C. Tauber. 1955. Spatial Price Equilibrium Models of the Livestock Feed Economy, *American Economic Review* 45: 584-608.

Frank, M. and T. Stengos. 1989. Measuring the Strangeness of Gold and Silver Rates of Return, *Review of Economic Studies*, 56:553-567.

Friesz, T.L., R.L. Tobin, T.E. Smith, and P.T. Harker. 1983. A Nonlinear Complementarity Formulation and Solution Procedure for the General Derived Demand Network Equilibrium Problem, *Journal of Regional Science,* 23: 337-61.

Fritsch, B., R. Condon, and B. Saugy. 1977. The Use of Input-Output Techniques in an Energy Related Model, *In* G. Bruckmann, (ed.) *Input-Output Approaches in Global Modeling*, New York: Pergamon Press.

Fuss, M.A. 1980. The Derived Demand for Energy in the Presence of Supply Constraints, W.T. Ziemba, S.L. Schwarts and E. Koenigsberg, (eds.) *Energy Policy Modeling* Boston: Martinus Nijhoff, 65-85.

Gabr, M.M. and T. Subba Rao. 1981. The Estimation and Prediction of Subset Bilinear Time Series Models with Applications, *Journal of Time Series Analysis, 2:* 155-171.

Gass, S. 1977. Evaluation of Complex Models, *Computer and Operations Research*, 4: 206-215.

Gately, D. 1979, The Prospects of OPEC Five Years After 1973/74 (A World Energy Model Survey), *European Economic Review*, 2: 369-379.

Gaudet, G. and N.M. Hang. 1986. Théorie économique des ressources non renouvelables: quelques éléments de synthèse, Université de Laval, Quebec.

Gerlow, M.E., S.H. Irwin and T.R. Liu. 1993. Economic Evaluation of Commodity Price Forecasting Models, *International Journal of Forecasting*, 9: 387-397.

Ghosh, S., C.L. Gilbert and A.J. Hughes. 1987. Stabilizing Speculative Commodity Markets. Oxford: Clarendon

Gilbert, C.L. 1991. Optimal and Competitive Storage Rules: The Gustafson Problem Revisited, In W. Labys and J.B. Lesourd (eds.) *International Commodity Market Models*, London: Chapman-Hall.

Gilbert, R. 1978. Dominant Firm Pricing Policy in a Market for an Exhaustible Resource, *Bell Journal of Economics*, 9: 385-95.

Gilbert, R. 1979. Search Strategies and Private Incentives for Resource Exploration, In R. Pindyck.(ed) *Advances in the Economics of Energy and Resources*. New York: JAI Press, 149-70.

Giraud, P.N. 1995. The Equilibrium Price Range of Oil, *Energy Policy*, 23:35-49.

Goldman, N.L., and J. Gruhl. 1980. Assessing the ICF Coal and Electric Utilities Model, In S.I. Gass (ed) *Validation and Assessment Issues of Energy Models*, National Bureau of Standards Special Publication 569, US Government Printing Office, Washington, DC.

Gordon, R.L. 1979. *Economic Analysis of Coal Supply: An Assessment of Existing Studies*. EPRI Report EA-496 (3 Vols). Electric Power Research Institute, Palo Alto, CA.

Granger, C.W.J., 1980. *Forecasting in Business and Economics*. San Francisco: Holden Day.

Granger, C.W.J., and P. Newbold. 1977. *Forecasting Economic Time Series*, New York: Academic Press.

Granger, C.W.J. and T. Terasvirta. 1993. *Modeling Nonlinear Economic Relationships*, Oxford: Oxford University Press.

Greenburg, M.J. and J.M. Conway. 1989. The Long Wave: Oil Prices, Commodity Prices and Economic Behavior, *Energy Center Quarterly*, 2:1-7.

Greenberger, M., M.A. Crenson, and B.L. Crissey. 1976. *Models in the Policy Process: Public Decision Making in the Computer Era*. New York: Russel Sage Foundation.

Greenberger, M. 1983. *Caught Unawares: The Energy Decade in Retrospect*. Cambridge: Ballinger Publishing Co.

Griffin, J.M. 1974. The Effects of Higher Prices on Electricity Consumption, *Bell Journal of Economics*, 5: 515-639.

Griffin, J.M. 1979. *Energy Conservation in the OECD:1980 to 2000*. Cambridge, MA: Ballinger.

Griffin, J.M. 1976. Energy Input-Output Modeling Problems and Prospects, EPRI Report EA-298. Electric Power Research Institute, Palo Alto.

Griffin, J.M. and P.R. Gregory. 1976. An Intercountry Translog Model of Energy Substitution Responses, *American Economic Review*, 66: 845-57.

Griffin, J.M. 1991. Methodological Advances in Energy Modeling: 1970-1990, *Energy Journal*, 14: 111-124.

Groncki. P.J. and W. Marcuse. 1980. The Brookhaven Integrated Energy/Economy Modeling System and Its Use in Conservation Policy Analysis, *Energy Modeling Studies and Conservation*. Economic Commission for Europe, New York: Pergamon Press for the United Nations, 535-556.

Grubb, M. 1993. Policy Modeling for Climate Change: The Missing Models, *Energy Policy*, 21: 203-208.

Gupta, S. 1981. *World Zinc Industry*. Lexington, MA: Heath Lexington Books.

Hannon, B. 1983. A Comparison of Energy Use, *Resources and Energy, 139-153*.

Hanslow, K. 1993. Incorporating Risks in Computable General Equilibrium Models. ABARE - CP 93.25, Economic Modeling Bureau of Australia, Cairns.

Harmon, P., R. Maus and W. Morrissey. 1988. *Expert Systems Tools and Applications,* New York: John Wiley and Sons, pp. 83-85.

Harris, D.V.P. 1984. *Mineral Resources Appraisal: Mineral Endowment, Resources, and Potential Supply: Concepts, Methods, and Cases*. Oxford: Clarendon Press, Oxford Geological Science Series.

Harker, P. 1988. Dispersed Spatial Price Equilibrium, *Environment and Planning*, A 20: 353-68.

Hartman, R.S. 1978. *A Critical Review of Single Fuel and Interfuel Substitution Residential Energy Demand Models*, Energy Lab Report, MIT-EL-78-003, MIT Energy Lab, Cambridge, MA.

Hartman, R. S. 1979. Frontiers in Energy Demand Modeling, *Annual Review of Energy*, 4: 43-466.

Harvey, A.C., 1985 The Trends and Cycles in Macroeconomic Times Series, *Journal of Business and Economic Statistics*, 3:216-227

Harvie, C.H. and T. Vantloa. 1993. Long-term Relationships of Major Macro-Variables in a Resource-Related Economic Model of Australia: A Cointegration Analysis, *Energy Economics*, 15: 258-261.

Harvie, C.H., and A. Thaha. 1994. Oil Production and Macroeconomic Adjustments in the Indonesian Economy, *Energy Economics*, 16: 253-270.

Hashimoto, H. 1977. World Food Projection Models, Projections, and Policy evaluation, Ph.D. Thesis, Department of Economics, University of Illinois.

Hashimoto, H., and T. Sihsobhon. 1981. A World Iron and Steel Economy Model: The WISE model, *In World Bank Commodity Models*. Washington, D.C.: The World Bank, 1-46

Haurie, A., Smeers, G and G. Zaccour. 1996. Gas Contract Portfolio Management: Experiments with a Stochastic Programming Approach, In J.B. Lesourd, J. Percebois, and F. Valette, (eds.) *Models for Energy Policy*, New York: Routledge.

Hazila, M. and Kopp, R. 1984. Modeling Mineral Demand Elasticities, *Journal of Enviornmental Economics and Management*, 12:17-28.

Henderson, J.M. 1958. *The Efficiency of the Coal Industry: An Application of Linear Programming*. Cambridge, MA: Harvard University Press.

Heo, E. 1996. An Econometric Analysis of the Short Run Behavior of the U.S. Petroleum Refining Industry, Presented at the MEMS Annual Conference, Montreal.

Herendeen, R.A. 1973. *The Energy Cost of Goods and Services*, Report ORNL-NSF-EP-58, Oak Ridge National Laboratory, Oak Ridge, TN.

Herrara, A. O., and H. D. Scolnik. 1976. Catastrophe or New Society: A Latin American World Model, International Development Research Center, Ottawa.

Heuttner, D.A. 1976. Net Energy Analysis: An Economic Assessment, *Science*, 192: 101-104.

Hibbard, W.R. 1980. An Disaggregated Supply Model of the US Copper Industry Operating in an Aggregated World Supply/Demand System, *Materials and Society*, 4: 261-284.

Hibbard, W.R., Soyster, A.L. and R.S. Gates. 1982, Supply Prospects for the U.S. Copper Industry: Alternative Scenarios: Midas II Computer Model, *Materials and Society*, 6: 201-210.

Hibbard, W.R., Soyster, A.L. and M.A. Kelly. 1979. *An Engineering Econometric Model of the U.S. Aluminum Industry*. New York: Proceedings of the American Institute of Mining Engineers.

Hibbard, W.R., Soyster, A.L and R.S. Gates. 1980. A Disaggregated Supply Model of the U.S. Copper Industry Operating in an Aggregated World Supply/demand System, *Materials and Society*, 4: 261-84.

Higgs, P.J. 1987. A Three Sector Miniature ORANI Model, WP OP-63, Impact Project, Monash University.

Hinchy, M., Hnaslow, K. and B.S. Fisher. 1994. A Dynamic Game Approach to Greenhouse Policy, ABARE 94.2, Australian Bureau of Agricultural and Resource Economics, Canberra.

Hitch, C.V. 1977. *Modeling Energy - Economy Interactions*. Washington, DC: Resources for the Future.

Hnyilicza, E., and R. Pindyck, 1976. Pricing Policies for a Two-Part Exhaustible Resource Cartel: The Case of OPEC, *European Economic Review*, 8: 136-154.

Hoel, M. 1993. Emission Taxes in a Dynamic International Game of CO_2 Emission, In Rüddiger Pethig (ed.), *Conflict and Cooperation in Managing Environmental Resources*, New York:SpringerVerlag, 39-68.

Hoel, M. 1994. The Role of a Carbon Tax in Environmental and Fiscal Policy, Paper presented at the 50th Congress of the International Institute of Public Policy, In *Public Finance, Environment and Natural Resources*, International Institute of Public Policy Cambridge, MA, 22-25

Hogan, W.H. and J.P. Weyant, 1980, Combined Energy Models, Discussion Paper E80-02, Kennedy School of Government, Harvard University, Cambridge, MA.

Hoffman, K.C., and D.W. Jorgenson. 1977. Economic and Technological Models for Evaluation of Energy Policy, *Bell Journal of Economics*, 8: 444-466.

Hoffman, K.C., and D.O. Wood. 1976. Energy Systems Modeling and Forecasting, *Annual Review of Energy*, 2: 423-453.

Hoffman, K., P.J. Groncki, and W.L. Graves. 1979. Energy Policy Analysis Forum: World Oil Analysis, Working paper, Brookhaven National Laboratory, Brookhaven, New York.

Holloway, M.L. 1980. *Texan National Energy Project: An Experiment in Large-Scale Model Transfer and Evaluation*, Vols. 1-3, New York: Academic Press.

Hope, C. (ed.) 1993. Policy Modeling for Global Climate Change, *Energy Policy,* 21: 202-338.

Hourcade, J-C. 1993. Modeling Long-Run Scenarios: Methodology Lessons From a Prospective Study on a Low CO_2 Intensive Country, *Energy Policy,* 21: 309-326.

Hua, C., and F. Porell. 1979. A Critical Review of the Development of the Gravity Model, *International Regional Science Review,* 4: 97-126.

Hudson, E.A., and D.W. Jorgenson. 1974. U.S. Energy Policy and Economic Growth. 1975-2000, *Bell Journal of Economics,* 5: 461-514.

Hughes, B.B., and M.D. Mesarovic. 1978. Analysis of the WAES Scenarios Using the World Integrated Model, *Energy Policy,* 6: 129-139.

Hunt, L. and N. Manning. 1989. Energy Price- and Income-Elasticities of Demand: Some Estimates for the UK Using the Cointegration Procedure, *Scottish Journal of Political Economy,* 36: 183-93.

Huntington, H.G., J.P. Weyant, and J.L. Sweeney. 1980. Modeling for Insights, not Numbers: the Experiences of the Energy Modeling Forum, *Omega: The International Journal of Management Science,* 1: 449-462.

Huntington, H.G. 1993. OECD Oil Demand: Estimated Response Surfaces for Nine World Oil Models, *Energy Economics,* 15: 49-56.

ICF Incorporated. 1977. *Coal and Electric Utilities Model Documentation,* 2nd edn. ICF, Inc., Washington, D.C.

ICF Incorporated. 1988. *Global Macro-Energy Model,* ICF, Inc., Washington, D.C.

Irwin, C., and C.W. Yang. 1983. Iteration and Sensitivity for a Nonlinear Spatial Equilibrium Problem, *Lecture Notes in Pure and Applied Mathematics* 85: 91-107.

Jacques, J.K., Lesourd, J.B. and J.M. Ruiz (eds.) 1988. *Modern Applied Energy Conservation,* New York: John Wiley.

Johansen, S. 1988. Statistical Analysis of Cointegration Vectors, *Journal of Economic Dynamics and Control,* 12: 183-93.

Johansen, S. 1991. Estimation and Hypothesis Testing of Cointegration Vectors in Gaussian Vector Autoregressive Models, *Econometrica,* 59:1551-80.

Johnson, J. 1980. *Econometric Methods.* New York: McGraw-Hill.

Johnson, S.R., and G.C. Rausser. 1977. Systems Analysis and Simulation: A Survey of Applications in Agricultural and Resource Economics, In G.Judge (ed.) *A Survey of Agricultural Economic Literature, Vol 2.* Quantitative Methods in Agricultural Economics, Minneapolis: University of Minnesota Press, 157-304.

Johnes, C.T. 1993. A Single-Equation Study of U.S. Petroleum Consumption: The Role of Model Specification, *Southern Economic Journal* 59: 687-700.

Jorgenson, D.W. and P.J. Wilcoxen. 1989. Environmental Regulation and U.S. Economic Growth, Center for Energy Policy Research, MIT Energy Lab, Cambridge, MA.

Jorgenson, D.W. and P.J. Wilcoxen, 1991. *Global Change, Energy Prices, and U.S. Economic Growth*, Cambridge, MA: Harvard University Press.

Jung, Suh. 1982. An Investment Planning Model for the Refining and Petrochemical Industry in Korea, Ph.D. Thesis, University of Texas at Austin.

Judge, G.G., and T.D. Wallace. 1959. *Spatial Price Equilibrium Analyses of the Livestock Economy*, Technical Bulletin TD-78, Department of Agricultural Economics, Stillwater: Oklahoma State University.

Karamardian, S. 1971. Generalized Complementarity Problem, *Journal of Optimization Theory Application,* 8: 161-68.

Karamardian, S.1972. The Complementarity Problem.*Mathematical Programming* 2:107-29.

Kaufman, G.M. 1983. Oil and Gas Estimation of Undiscovered Resources, In M.A. Adelman (ed.) *Energy Resources in an Uncertain Future; Coal, Gas, Oil and Uranium Supply Forecasting,* Cambridge, MA: Ballinger.

Kendrick, D. 1967. *Programming Investment in the Process Industries: An Approach to Sectoral Planning.* Cambridge: MIT Press.

Kendrick, D., and A. Stoutjesdijk. 1978. *The Planning of Industrial Investment Programs: A Methodology.* Baltimore: Johns Hopkins University Press for the World Bank.

Kendrick, D., A. Meeraus, and J.S. Suh. 1981. *Oil Refinery Modeling with the GAMS Language,* Center for Energy Studies, Research Report 14, The University of Texas, Austin.

Kendrick, D., A. Meerus, and J. Alatorre. 1984. *The Planning of Investment Programs in the Steel Industry.* Baltimore: Johns Hopkins University Press for the World Bank.

Kennedy, M., 1974, An Economic Model of the World Oil Market, *Bell Journal of Economics,* 5: 540-577.

King, T.B. and B.J. Reddy. 1981. *Analysis of the Effect of Price Increases on the Demand for Manganese,* Board on Mineral and Energy Resources, Committee on Nonfuel Mineral Demand Relationships, National Academy of Sciences, Washington, DC.

Kim, T-Y and S-R Kae. 1993. An Integrated Energy Policy for Korea, *Energy Policy,* 21: 1001-1010.

Klein, L.R. (ed.). 1992. *Modeling Global Change.* Tokyo: United Nations University Press.

Kocagil, A.E. 1997. Does Futures Speculation Stabilize Spot Prices? *Applied Financial Economics,* 7:115-125.

Kolk, D. 1983. The Regional Employment Impact of Rapidly Escalating Energy Costs, *Energy Economics.*5:67-78

Kolstad, C.D. 1982. Noncompetitve Analysis of the World Coal Market, Working paper. Energy Laboratory, Los Alamos.

Kolstad, C.D. and A. Mathiesen. 1991. Computing Equilibria in Imperfectly Competitive Commodity Markets, In W.C. Labys and J.B. Lesourd (eds.) *International Commodity Market Models*, London: Chapman & Hall.

Kolstad, C., D. Abbey, and R. Bivins. 1983. Modeling International Steam Coal Trade. Working Paper LA-9661-MS, Los Alamos, N.M.: Los Alamos National Laboratory.

Kouassi, E, W.C. Labys, and D. Colyer. 1996. Structural Time Series Approach to Forecasting Commodity Prices. Working Paper, Natural Resource Economics Program, West Virginia University, Morgantown.

Kovisars, L. 1982. Uranium 1981-2000, MET Research, Dallas, TX.

Kovisars, L. 1982. Molybdenum 1982-2000, MET Research, Dallas, TX.

Kovisars, L. 1975. *Copper Trade Flow Model.* SRI Project MED 3742-74, Stanford Research Institute, Stanford.

Kovisars, L. 1976. *World Production Consumption and Trade in Zinc - An LP Model*, U.S. Bureau of Mines Contract Report J-0166003, Stanford Research Institute, Stanford.

Kress, A., Robinson, D., and K. Ellis. 1992. Comparison of the Structure of International Oil Models, In *International Oil Supplies and Demands*, Vol. 2, Energy Modeling Forum, Stanford, CA.

Krueger, P.K. 1976. Modeling Future Requirements for Metals and Minerals, *Proceedings of the XIV Symposium of the Council for the Application of Computers and Mathematics in the Minerals Industry.* University Park, PA: Pennsylvania State University.

Kuenne, R.E. 1980. Modeling the OPEC Cartel with Crippled Optimization Techniques, Paper presented at the Conference on World and World Region Energy Studies, International Congress of Arts and Sciences, Harvard University, Cambridge, MA

Kuh, E., and D.O. Wood. 1979. Independent Assessment of Energy Policy Models• Report EA-1071, Electric Power Research Institute, Palo Alto, CA.

Kuhn, N.W., and A.W. Tucker. 1951. Nonlinear programming, In J. Neyman (ed.) *Proceedings of the Second Berkeley Symposium on Mathematical Statistics and Probability*, Berkeley:University of California Press.

Kumar, K. 1986. On the Identification of Some Bilinear Time Series Models, *Journal of Time Series Analysis*, 7:117-122.

Kwang, H.K. 1981. An Investment Programming Model in the Electric Power Industry. Ph.D. Thesis, University of Texas at Austin.

Labys, W.C. 1973. *Dynamic Commodity Models: Specification, Estimation and Simulation.* Lexington, MA: Heath Lexington Books.

Labys, W.C. (ed.) 1975. *Quantitative Models of Commodity Markets.* Cambridge, MA: Ballinger.

Labys, W.C. 1976. Naivety and Optimality on Commodity Price Forecasting. In W. Driehuis *(ed.) Primary Commodity Prices: Analysis and Forecasting.* Rotterdam: Rotterdam University Press.

Labys, W.C. 1977. Minerals Commodity Modeling: The State of the Art, *Proceedings of the Mineral Economic Symposium on Minerals Policies in Transition.* Washington, DC: Council of Economics of the AIME, 80-106.

Labys, W.C. 1978. Commodity Markets and Models: The Range of Experience, *In* F. G. Adams, (ed.) *Stabilizing World Commodity Markets: Analysis, Practice and Policy,* Lexington, MA: Heath Lexington Books.

Labys, W.C. 1980a. *Market Structure, Bargaining Power and Resource Price Formation.* Lexington, MA: Heath Lexington Books.

Labys, W.C. 1980b. A Model of Disequilbrium Adjustments in the Copper Market, *Materials and Society,* 4: 153-164.

.Labys, W.C. 1981. A General Disequilibrium Model of Commodity Market Adjustments, NSF Project Report DAR 78-08810, Department of Mineral and Energy Resource Economics, West Virginia University, Morgantown.

Labys, W.C. 1982a. Measuring the Validity and Performance of Energy Models, *Energy Economics,* 4: 159-168.

Labys, W.C. 1982b. A Critical Review of International Energy Modeling Methodologies, Energy Lab Working Paper, 82-034WP, MIT Energy Lab, Cambridge, MA

Labys, W.C. 1987. *Commodity Markets and Models: An International Bibliography.* London: Gower Publishing Co.

Labys, W.C. 1993. Model Validation, In J.P. Weyant and T.A. Kuczmowski. (eds.) *Systems Modeling Handbook,* New York: Pergamon Press.

Labys, W.C. 1996. *Modeling Methods Important for the Polish Mineral and Energy Industries.* Mineral and Energy Economic Research Institute, Polish Academy of Sciences, Krakow.

Labys, W.C., Afrasiabi, A and M. Moallem. 1991. Spectral Analysis of Stock Adjustments in Mineral Markets, In W.C. Labys and J.B. Lesourd (eds.) *International Commodity Market Models*. London: Chapman & Hall.

Labys, W.C., Elliott, C. and H. Rees. 1971. Copper Price Behavior and the London Metal Exchange, *Applied Economics,* 3:99-113.

Labys, W.C., F.R. Field, and J. Clark. 1985. Mineral modeling, In W. Vogely (ed.) *Economics of the Mineral Industry.* New York: American Institute of Mining Engineers.

Labys, W.C. and C.W.J. Granger. 1970. *Speculation, Hedging and Commodity Price Forecasts.* Lexington: Heath Lexington Books.

Labys, W.C., and E. Kouassi. 1996. Structural Time Series Modeling of Commodity Price Cycles, Research Paper 9602, Regional Research Institute, West Virginia University, Morgantown.

Labys, W.C. and J.B. Lesourd (in association with O. Guvenen). 1991. *International Commodity Market Models.* London: Chapman & Hall.

Labys, W. C., J-B Lesourd, and J. K. Jacques. 1988. Forecasting for Energy Management. In . J.K. Jacques, J-B Lesourd, and J-M Ruiz *(eds.) Modern Applied Energy Conservation.* New York and London: Ellis Harwood/John Wiley.

Labys, W.C., Badillo, D. and J.B. Lesourd. 1995. The Cyclical Behavior of Individual Commodity Price Series, Document de Travail 95B03, GREQAM, Universities Aix-Marseille II, III, Marseille

Labys, W. C., J. B. Lesourd, and N. Uri. 1990. New Horizons in Commodity Modeling, In W. C. Labys, and J. B. Lesourd *(eds.) New Directions in International Commodity Market Modeling,* London: Chapman & Hall.

Labys, W.C. and M. Lord. 1992. Inventory and Equilibrium Adjustments in International Commodity Markets: a Multi-Cointegration Approach, *Applied Economics,* 24:77-84.

Labys, W.C., and A. Maizels. 1993. Impact of Commodity Price Fluctuations on the Developed Economies, *Journal of Policy Modeling,* 15: 335-352.

Labys, W.C., V. Murcia and M. Terraza. 1996. Modelling the Petroleum Spot Market: a Vector Autoregressive Approach• In J.B. Lesourd, J. Percebois, and F. Valette (eds.) *Models for Energy Policy,* New York: Routledge.

Labys, W C., S. Paik, and A. M. Liebenthal. 1979. An Econometric Simulation Model of the U.S. Steam Coal Market, *Energy Economics,* 1: 19-26.

Labys, W.C., and P.K. Pollak. 1984. *Commodity Models for Forecasting and Policy Analysis.* London: Croom-Helm.

Labys, W.C., T. Takayama, and N. Uri. 1989. *Quantitative Methods for Market Oriented Economic Analysis Over Space and Time.* London: Gower Publishing Co.

Labys, W.C., and J.P. Weyant. 1990. Integrated models, In J.P. Weyant and T.A. Kuczmowski (eds.) *Engineering-economic Modeling: Energy Systems.* New York: Pergamon Press.

Labys, W C., and D.O. Wood. 1985. Energy modeling, In W. Vogely (ed.). *Economics of the Mineral Industry.* New York: American Institute of Mining Engineers.

Labys, W.C., and C.W. Yang. 1980. A Quadratic Programming Model of the Appalachian Steam Coal Market, *Energy Economics* 2: 86-95.

Labys, W.C., and C.W. Yang. 1991. Advances in the Spatial Modeling of International Mineral and Energy Issues, *International Regional Science Review* 14: 61-94.

Labys, W.C. and C.W. Yang. 1996. Le Chatelier Principle and the Allocation Sensitivity of Spatial Commodity Models. In J.C. van der Bergh, P. Nijkamp and P. Rietveld (eds) *Recent Advances in Spatial Equilibrium Modeling: Methodology and Application (essays in honor of Takashi Takayama),* Berlin: Springer-Verlag.

Labys, W.C. and C.W. Yang. 1997. Spatial Price Equilibrium as the Core to Spatial Commodity Modeling. *Papers in Regional Science* 76: 199-228.

Langston, V.C. 1983. An Investment Model for the U.S. Gulf Coast Refining Petrochemical Complex, Ph.D. thesis, University of Texas at Austin.

Larson, D.F. 1994. Copper and the Negative Price of Storage, PRWP. 1282, International Economics Department, World Bank, Washington, D.C.

Lawphongpanich, S., and D.W. Hearn. 1984. Simplicial Decomposition of the Asymmetric Assignment Problem, *Transportation Research,* 18B: 123-33.

Lau, L.J. 1982. The Measurement of Raw Material Inputs, V.K. Smith and J.V. Krutilla, (eds.) In *Explorations in Natural Resource Economics,* Baltimore, MD: John Hopkins University Press for Resources for the Future, Inc., 167-200.

Le Goff, P. 1979. *Energétique industrielle,* vol. 2, Paris: Laviosier,.

Lemke, C.E. 1965. Bimatrix Equilibrium Points and Mathematical Programming, *Management Science,* 11: 681-89.

Lemke, C.E., and J.T. Howson, Jr. 1964. Equilibrium Points of Bimatrix Games, *Journal of the Society of Industrial Application of Mathematics,* 12: 413-23.

Leontief, W. W. 1951. *The Structure of the American Economy 1919-1939.* New York: Oxford University Press.

Leontif, W., Koo, J. Nasar S. and I. Sohn. 1983. *The Future of Nonfuel Minerals in the U.S. and World Economy: Input-Output Projections 1980-2030.* Lexington, MA: Heath Lexington Books.

Lesourd, J.B., Percibois, J., and F. Valette (eds). 1996. *Models for Energy Policy.* New York: Routledge.

Lesourd, J.B. 1984. *Energie et substitutions entre facteurs de production*, Grenoble: Presses Universitaires de France.

Lesourd, J.B. and A. Consonni. 1984, Production Models of Energy Efficiency, In A. Reis, J.L. Puebe, I. Smith, and K. Stephan (eds), *Energy Economics and Management in Industry.* London: Pergamon Press.

Lesourd, J.B. and Y. Gousty (eds.). 1981. *Le management de lénergie.* Paris: Masson.

Lev, B. 1983. *Energy Model Studies.* Amsterdam: North Holland

Lev, B., F.H., Bloom, J.A. and A.S. Gleit. 1984. *Analytical Techniques for Energy Planning.* Amsterdam: North Holland.

Libbin, J.D., and M.D. Boehji. 1977. Interregional Structure of the U.S. Coal Industry, *American Journal of Agricultural Economics* 59: 456-66.

Lin, S. M. 1991. Computable General Equilibrium Models for U.S. Nonfuel Minerals Policy, Ph.D. Thesis, West Virginia University, Morgantown.

Linage, D.R. 1974. *Energy Policy Evaluation.* Lexington, MA: Health Lexington Books.

Linneman, H.H. 1966. *An Econometric Study of International Trade Flows.* Amsterdam: North-Holland.

Linneman, J. 1976. MOIRA: Model of International Relations in Agriculture - The Energy Sector, Working Paper, Institute for Economic and Social Research, Free University, Amsterdam.

Lofting, E. 1979. *The Input-Output Structure of U.S. Mineral Industries.* US Bureau of Mines, Washington, DC.

Lord, M.J. 1991. *Imperfect Competition and International Commodity Trade.* Oxford: Clarendon Press.

MacAvoy, P., and R. Pindyck. Alternative Regulatory Policies for Dealing with the Natural Gas Shortage, *Bell Journal of Economics,* 4: 454-98.

MacAvoy, P.W., and R.S. Pindyck. 1975. *The Economics of the Natural Gas Shortage 1960-1980*, Amsterdam: North-Holland.

MacFadden, A.J. 1993. OPEC and Cheating: Revisiting the Kinked Demand Curve, *Energy Policy*, 21: 858-867.

Mackey, M.C. 1989. Commodity Price Fluctuations: Price Dependent Delays and Nonlinearities as Explanatory Factors, *Journal of Economic Theory*, 48:497-509.

Madlener, R. 1996. Econometric Methodologies of Estimating Household Energy Demand: A Survey, *The Journal of Energy Literature*, 2:3-32.

Malinvaud, E. 1978. *Statistical Methods of Econometrics*, Revised Edition, Chicago: Rand McNally.

Manne, A.S. 1977. ETA-Macro: A Model Energy-Economy Interactions, In C. Hitch (ed.) *Modeling Energy-Economy Interactions*. Washington, DC: Resources for the Future.

Manne, A.S., and H.M. Markowitz, (eds). 1963. *Studies in Process Analysis*. New York: John Wiley & Sons.

Manne, A.S. and R.G. Richels. 1992. *Buying Greenhouse Insurance: The Economic Cost of CO_2 Emission Limits*. Cambridge, MA: MIT Press.

Manne, A.S., Mendelsalin, R., and R. Richards. 1995. NERGE. A Model for Evaluating Regional and Global Effects of GH6 Reduction Policies, *Energy Policy,* 23: 17-34.

Marschak, T.A. 1963. A Spatial Model of U.S. Petroleum Refining, In A. S. Manne and H. M. Markowitz. (eds.) *Studies in Process Analysis*, New York: John Wiley & Sons.

McDonald, R., and D. Seigel. 1986. The Value of Waiting to Invest, *Quarterly Journal of Economics,* 707-727.

McKnight, R.T. 1988. An Empirical Methodology for the Valuation of Risky Gold Cashflows, In C.O. Brawner, (ed.) *Gold Mining 88.* Littleton Co.: Society of Mining Engineers.

McLeod, A.I. and W.K. Li. 1983. Diagnostic Checking of ARMA Time Series Models Using Squared-residuals Autocorrelations, *Journal of Time Series Analysis*, 4:269-273.

Meadows, D.L. 1970. *Dynamics of Commodity Production Cycles*. Cambridge: Wright-Allen Press.

Meirer, P. 1985a. Energy Planning in Developing Countries: The Role of Microcomputers, *Natural Resources Forum*, 9: 41-52.

Meirer, P. 1985b. Energy Systems Analysis for Developing Countries, *Lecture Notes in Mathematical Systems* 222, New York: Springer-Verlag.

Meirer, P. 1986. *Energy Planning in Developing Countries: An Introduction to Analytical Methods*. Boulder, CO: Westview Press.

Meirer, P. and V. Mubayi. 1981. A Linear Programming Framework for Analysis of Energy-economic Interactions in Developing Countries, *European Journal of Operations Research*, 13:41-9.

Miernyk. W., 1978. *Regional Impacts of Rising Energy Prices,* Cambridge: Ballinger.

Miranda, M.J., and J.W. Glauber. 1991. The Spatial-Temporal Price Equilibrium Model Under Uncertainty, In W.C. Labys and J. B. Lesourd (eds.), *International Commodity Market Models*, London: Chapman & Hall.

Mitchell, B.M., R.E. Park, and F. Labrune. 1986. *Projecting the Demand for Electricity: A Survey and Forecast*, Series 3312-PSSP, Rand Corporation. Santa Monica, CA

Moore, G.H. 1980. *Business Cycles, Inflation and Forecasting.* Cambridge: National Bureau of Economic Research.

Moore, G.H. 1988. Inflation Cycles and Metals Prices, *Mineral Processing and Extractive Metallurgy Review*, 3:95-104.

Mork, K.A., Mysen, H.T. and O. Olsen. 1990. Business Cycles and Oil Price Fluctuations In O. Bjerkholt, O. Olsen and J. Vislie (eds.), *Recent Modeling Approaches in Applied Energy Economics.* London: Chapman & Hall.

Moroney, J.R. (ed.) 1984. *Econometric Models of the Demand for Energy.* Greenwich, CN: JAI Press.

Moroney, J.R. 1982. *Formal Energy and Resource Models.* Greenwich, CN: JAI Press.

Moroney, J. and D. Bremmer. 1987. An Integrated Regional Petroleum Model, In J. Moroney (ed.), *Advances in the Economics of Energy Resources.* Greenwich, CT: JAI Press, 187-220.

Morrison, W., and P. Smith. 1974. Nonsurvey Input-Output Techniques at the Small Area Level, *Journal of Regional Science*, 14, No. 1.

Morrison, W.E., and C.L. Readling. 1968. *An Energy Model for the United States: Featuring Energy Balances for the Years 1947 to 1965 and Projections and Forecasts to the Years 1980 and 2000*, IC 8384, U.S. Bureau of Mines, Washington, DC.

Munasinghe, M. 1990. *Energy Analysis and Policy.* London: Butterworth.

Munasinghe, M. 1979. *The Economics of Power System Reliability and Planning.* Baltimore: Johns Hopkins University Press.

Munasinghe, M. 1980. Integrated National Energy Planning in Developing Countries, *Natural Resources Forum*, 4: 359-73.

Munasinghe, M. 1981. Principles of Modern Electricity Pricing, *Proceeding of International Electric Engineers*, 69:332-48.

Munasinghe, M. 1986. Practical Applications of Integrated Natural Energy Planning (INEP) Using Microcomputers, *Natural Resources Forum*, 10:17-38.

Munasinghe, M. 1988. Integrated National Energy Planning and Management: Methodology and Application to Sri Lanka, Technical Paper 86, World Bank, Washington, DC.

Munasinghe, M. and P. Meier. 1993. *Energy Policy Analysis and Modeling.* Cambridge: Cambridge University Press.

Munasinghe, M. and P. Meier. 1985. Hierarchical Modeling for Integrated National Energy Planning: Microcomputer Implementation, In M. Munasinghe, M. Dow, and J. Fritz (eds.), *Microcomputers for Development: Issues and Policy,* CINTEC and National Academy of Sciences, Colombo and Washington, DC.

Munasinghe, M. and J.J. Warford. 1981. *Electricity Pricing.* Baltimore: Johns Hopkins University Press.

Murphy, F.H. 1980. The Structure and Solution of the Project Independence Evaluation System, U. S. Energy Information Administration, Washington, DC.

Murphy, F.H., Sanders, R.C., Shaw, S.H., and R.L. Thrasher. 1980, Modeling Natural Gas Regulatory Proposals Using the Project Independence Evaluation System, *Operations Research,* 29: 876-902.

Murphy, F.H., Sherali, H.D. and A.L. Soyster. 1980. A Mathematical Programming Approach for Determining Oligopolistic Market Equilibrium, Working Paper. Virginia Polytechnic Institute and State University, Blacksburg.

Murty, K.G. 1972. On the Number of Solutions of the Complementarity Problem and Spanning Properties of Complementary Cones, *Linear Algebra and Its Applications,* 5: 65-108.

Nagurney, A. 1987. Competive Equilibrium Problems, Variational Inequalities, and Regional Science. *Journal of Regional Science, 27:503-18*

Nash, J.F. 1951. Non-Cooperative Games, *Annals of Mathematics,* 54: 286-295.

Nash, J.F. 1953. Two-Person Cooperative Games, *Econometrica,* 21: 128-140.

National Academy of Sciences, (NAS), 1972. Mutual Substitutability of Aluminum and Copper, NMAB 286. National Materials Advisory Board, National Academy of Sciences, Washington, D.C.

National Academy of Sciences (NAS). 1982. *Mineral Demand Modeling.* Committee on Nonfuel Mineral Demand Relationships, Washington, D.C.: National Academy Press.

National Petroleum Council, 1974. *Emergency Preparedness for Interruption of Petroleum Imports into the United States.* Washington, D.C.: National Petroleum Council.

Nelson, J.P. 1970. An Interregional Recursive Programming Model of the U.S. Iron and Steel Industry 1947-67, Ph.D. Thesis, University of Wisconsin, Madison.

Nelson, C.R. 1973. *Applied Time Series Analysis for Managerial Forecasting.* San Francisco: Holden Day.

Newberry, D.M. and J.E. Stiglitz. 1981. *The Theory of Commodity Price Stabilization.* Clarendon: Oxford University Press.

Newcomb, R.T., and J. Fan. 1980. *Coal Market Analysis Issues.* Report EA-1575, Electric Power Research Institute, Palo Alto, CA.

Newcomb, R.T., Reynolds, S.S., and T.A. Masburch. 1990. Changing Patterns of Investment Decision-Making in World Aluminum, *Resources and Energy,* 11: 261-97.

Niedercorn, J.H., and J.D. Moorehead. 1974. The Commodity Flow Gravity Model: A Theoretical Reassessment, *Regional and Urban Economics,* 4: 69-75.

Nielssen, T. and A. Nystand. 1986. Optimum Exploration and Extraction in a Petroleum Basin, *Resources and Energy,* 8:219-30.

Nziramasanga, M. and C. Obideguri. 1981. Primary Commodity Price Fluctuations in Developing Countries: An Econometric Model of Copper and Zambia, *Journal of Developing Countries, 9.*

Ogawa, K. 1982. A New Approach to Econometric Modeling: A World Copper Model. NSF Report, Department of Economics, University of Pennsylvania, Philadelphia.

Ogawa, K. 1983. A Theoretical Appraisal of Price Stabilization Policy Under Alternative Expectational Schemes, Discussion Paper 9, Faculty of Economics, Kyobe University, Japan:

O'Leary, D.E. 1987. Validation of Expert Systems with Applications to Auditing and Accounting Expert Systems, *Decision Sciences,* 18: 468-486.

Opyrchal, A.M. and K.L. Wang. 1981. *Economic Significance of the Florida Phosphate Industry: An Input-Output Analysis,* IC8850, U.S. Bureau of Mines, Washington, D.C.

Organization for Economic Cooperation and Development. 1977. *World Energy Outlook* Paris: OECD

Pachauri, R.K. and L. Srivastava. 1988. Integrated Energy Planning in India: A Modeling Approach, *Energy Journal,* 9:35-48.

Padilla, V.P. 1992. Modeling Oil Exploration, In T. Sterner (ed.) *International Energy Economics.* London: Chapman & Hall.

Pakravan, K. 1977. A Model of Oil Production Development and Exploration• *Journal of Energy and Development,* 3: 143-53.

Palm, S.K., Parson, N.D. and JR. J.A. Read. 1986. Option Pricing: A New Approach to Mine Valuation, *CAM Bulletin,* May: 61-66.

Park, Se-Hark. 1982. An Input-Output Framework for Analyzing Energy Consumption, *Energy Economics,* 4: 105-110.

Peck, S.C., and T.J. Teisberg. 1993. CO_2 Emissions Control: Comparing Policy Instruments, *Energy Policy*, 21: 222-230.

Percebois, J and F. Valette. 1996. Modeling the European Gas Market: A Comparison of Several Scenario (SOSIE-GAZ), In J.B. Lesourd, J. Percebois, and F. Valette, (eds.), *Models for Energy Policy*, New York: Routledge.

Perron, P. 1989. The Great Crash, the Oil Shock and the Unit Root Hypothesis, *Econometrica*, 57:1361-1402.

Perroni, C. and R.M. Wigle. 1994. International Trade and Environmental Quality: How Important are the Linkages? *Canadian Journal of Economics,* 54: 325-38.

Peterson, F. 1978. A Model of Mining and Exploring for Exhaustible Resources, *Journal of Environmental Economics and Management*, 5:236-51.

Petruccelli, J.D. and N. Davies. 1986. A Portmanteau Test for Self Exciting Threshold Autoregressive-type Non-linearity in Time Series, *Biometrika,* 73:687-694.

Phlips, L. 1974. *Applied Consumption Analysis.* North-Holland, Amsterdam.

Pindyck, R.S. 1978a. The Optimal Exploration and Production of Nonrenewable Resources, *Journal of Political Economy*, 86:841-61.

Pindyck, R.S. 1978b. Gains to Producers from the Cartelization of Exhaustible Resources, *Review of Economics and Statistics,* 60: 238-51.

Pindyck, R.S. 1979. *The Structure of World Energy Demand*, Cambridge and London: MIT Press.

Pindyck, R.S. 1980. Uncertainty and Exhaustible Resources Markets, *Journal of Political Economy*, 88:1203-25.

Pindyck, R.S. 1994, Inventories and the Short-Run Dynamics of Commodity Prices, *RAND Journal of Economics*, 25:141-159.

Pindyck, R.S. and D. Rubinfield. 1997. *Econometric Models and Economic Forecasts*, New York: McGraw-Hill Publishing Company.

Pindyck, R.S. and J.J. Rotemberg 1990. The Excess Co-Movement of Commodity Prices, WP 2671, National Bureau of Economic Research, Cambridge, MA.

Plourde, A. and G.C. Watkins. 1995. How Volatile are Crude Oil Prices? *OPEC Review*, 19: 10-23.

Poyhonen, P. 1963. A Tentative Model of Trade Between Countries, *Weltwirtschafliches Archives*, 90: 93-100.

Preston, R. 1975. The Wharton Long-Term Model: Input-Output Within the Context of a Macro Forecasting Model, *International Economic Review*, 3-19.

Price, J.F. 1984. Coal Supply Models: The State of the Art, In B.Lev, J.A. Bloom. F.H. Murphy and A.S. Gleit (eds.), *Analytic Techniques for Energy Planning.* Amsterdam: North-Holland

Propoi, A., and I. Simin. 1981. Dynamic Linear Programming Models of Energy, Resource and Economic Development Systems, RR-81-14, International Institute for Applied Systems Analysis, Laxenburg.

Provenzano, G. 1989. Alternative National Policies and the Location Patterns of Energy-related Facilities, In W. C. Labys, T. Takayama, and N. Uri (eds.), *Quantitative Methods for Market-oriented Economic Analysis Over Space and Time.* London: Gower Publishing Co.

Puttock, G.D. and M. Sabourin. 1992. International Trade in Forest Products: An Overview, International Agricultural Trade Research Consortium, St. Petersburg, Fl.

Quick, A.N. and J.R. Schuler. 1988. An Expert System for Strategic Planning, *Oil and Gas Journal,* 8: 75-79.

Quirk, J., K. Terasawa, and D. Whipple. 1982. *Coal Models and their Use in Government Planning.* New York: Praeger Publishers.

Ray, W.H. and J. Szekely, 1973. *Process Optimization: With Application in Metallurgy and Chemical Engineering.* New York: John Wiley & Sons.

Reardon, W.A. 1972. *An Input/Output Analysis of Energy Use Changes from 1947 to 1958 and 1958 to 1963.* Office of Science and Technology, Executive Office of the President, Washington, D.C.: US Government Printing Office.

Richard, D., 1977, A Dynamic Model of the World Copper Industry, Working Paper, International Monetary Fund, Washington, DC.

Ridker, R.G. and W.D. Watson, 1980. *To Choose a Future,* Baltimore: Johns Hopkins University Press.

Ringwood, J.V., Austin, P.C. and W. Montieth. 1993. Forecasting Weekly Energy Consumption, *Energy Economics,* 15: 285-296.

Roehner, B.M. 1995. *Theory of Markets: Trade and Space-time Patterns of Price Fluctuations.* Berlin: Springer-Verlag.

Rosa, T.P., and M.T. Tolmasquin. 1993. An Analytical Model to Compare Energy-efficiency Indices and CO_2 Emissions in Developed and Developing Countries, *Energy Policy,* 21: 276-283.

Rose, A. 1989. *Regional Analysis of Energy Demands.* New York: Oxford University Press.

Rose, A. 1983. Technological Change and Input-Output Analysis: An Appraisal, Working Paper, Mineral and Energy Resource Economics Program, West Virginia University.

Rose, A., and D. Kolk. 1987. *Forecasting Natural Gas Demand in a Changing World.* Greenwich, CN: JAI Press.

Rose, A. 1974. A Dynamic Interindustry Model for the Economic Analysis of Pollution Abatement, *Environment and Planning*, A 6:361-338.

Rose, A. 1977. A Simulation Model for the Economic Assessment of Alternative Air Pollution Regulations, *Journal of Regional Science*, 17:327-344.

Rose, A. 1983. Modeling the Macroeconomic Impact of Air Pollution Abatement, *Journal of Regional Science*, 23:441-459.

Rose, A., and Miernyk, W. 1987. Input-Output Analysis: The First Fifty Years. *Journal of Policy Modeling.*

Roy, J. R. 1987. An Alternative Information Theory Approach for Modeling Spatial Interaction, *Environment and Planning, A* 19: 385-94.

Runge, C.F. 1994. *Environmental Effects of Trade in the Agricultural Sector: A Case Study.* Working Paper P92-1. Center for International Food and Agricultural Policy, St. Paul, MN

Ruth, M. 1993. *Integrating Economics, Ecology and Thermodynamics,* Dordrecht: Kluwer Academic Publishers.

Ruth, M. 1995a. Thermodynamic Implications for Natural Resource Extraction and Technical Change in U.S. Copper Mining, *Environmental and Resource Economics*, 6:187-206.

Ruth, M. 1995b. Technology Change in U.S. Iron and Steel Production: Implications for Material and Energy Use and CO_2 Emissions, *Resources Policy*, 21:199-214.

Ruth, M. and B. Hannon. 1997. *Modeling Dynamic Economic Systems*, New York: Springer-Verlag.

Ryan, D. and J. Livernois. 1985. Testing for non-Jointness in Oil and Gas Exploration: A Variable Profit Function Approach? Discussion Paper 85-6, University of Calgary, Calgary.

Salant, S., Sanghi, A. and M. Wagner. 1979. Imperfect Competition in the International Energy Market: A Computerized Nash-Cournot Model, ICF Report, U.S. Department of Energy, Washington, DC.

Salant, S., F. Miercort, Sanghvi, A., and M. Wagner. 1981. *Imperfect Competition in the International Energy Market.* Lexington, MA: Heath Lexington Books.

Sakarat, S. 1996. Evaluation of Mineral Projects Using Simulation and Expert Systems, Ph.D. Thesis, West Virginia University, Morgantown.

Samoulidis, J.E. and Mitropoulos, C.S. 1982. Energy-Economy Models: a Survey, *European Journal of Operational Research*, 11: 222-32.

Samuelson, P.A., 1952, Spatial Price Equilibrium and Linear Programming, *American Economic Review*, 42: 283-303.

Savage, I.R., and K.W. Deutsch. 1960. A Statistical Model of the Gross Analysis of Transactions Flows, *Econometrica*, 28: 551-72.

Scarf, H., 1973, *The Computation of Economic Equilibria*. New Haven: Yale University Press.

Scarfe, B.L. and E. Rilkoff. 1984. Financing Oil and Gas Exploration and Development Activity, Working Paper 274, Economic Council of Canada.

Schinzinger, R. 1974. Integer Programming Solutions to Problems in Electric Energy Systems, In D.R. Limaye *(ed.)*, *Energy Policy Evaluation.*, Lexington, MA: Heath Lexington Books.

Schlottman, A., and R.A. Watson. 1989. Air Quality Standards, Coal Conversion, and the Steam-electric coal market, In W.C. Labys, T. Takayama, and N. Uri (eds.) *Quantitative Methods for Market-oriented Economic Analysis Over Space and Time*, London: Gower Publishing Co.

Schurr, S. 1960. *Energy Modeling*. Washington, DC: Resources For the Future.

Schweppe, F.C., M. Caramanis, R. Tabors, and R. Bohn. 1988. *Spot Pricing of Electricity*. New York: Kluwer Publishing Co.

Serletis, A. 1994. A Cointegration Analysis of Petroleum Futures Prices, *Energy Economics*, 16: 93-98.

Shlyakhter, A. 1994. Quantifying the Credibility of Energy Projections from Past Trends: the U.S. Energy Sector, *Energy Economics, 16*: 119-130.

Siddayao, C.M. 1980. The Supply of Petroleum Reserves in South-east Asia, *In Economic Implications of Evolving Property Rights Arrangements*. Oxford: Oxford University Press.

Sims, C. 1980. Macroeconomics and Reality. *Econometrica*, 48:1-48.

Slade, M. 1981. Cycles in Natural Resources Commodity Prices: An Analysis in the Frequency Domain, *Journal of Environment Economics and Management*, 9: 138-148.

Smith, G.W., and Schenk, G. 1979, The International Tin Agreement: A Reassessment, *Economic Journal*.

Smith, C.B. 1976. *Efficient Electricity Use*. New York: Pergamon Press.

Sowell, F. 1992. Modeling Long-Run Behavior With the Fractional ARIMA Model, *Journal of Monetary Economics*, 29:277-302.

Soyster, A., and H.D. Sherali, 1981, On the Influence of Market Structure in Modeling the U.S. Copper Industry *Management Science:* 381-388.

Sparrow, F. and A. Soyster, 1980, Process Models of Minerals Industries, *Proceedings of the Council of Economics of the AIME,* New York, NY: American Institute of Mining, Metallurgical and Petroleum Engineers, 93-101.

Steenblik, R. P. 1985. Issues in Modeling International Coal Supply, EURICES Paper 85-2. Centre for International Energy Studies, Erasmus University, Rotterdam.

Steenblick, R.P. 1986. International Coal Resource Assessment, Working Paper, Department of Geography, London School of Economics.

Steenblick, R.P. 1992. Modeling the Long Run Supply of Coal, In T. Sterner (ed.) *International Energy Economics.* London: Chapman & Hall.

Stern, D.I. 1993. Energy and Economic Growth in the United States: A Multivariate Approach, *Energy Economics*, 15: 137-150.

Sterner, T. 1990. *The Pricing of and Demand for Gasoline.* Swedish Transport Research Board, Stockholm.

Sterner, T. (ed.) 1992. *International Energy Economics.* London: Chapman & Hall.

Sterner, T. and C. Dahl. 1992. Modeling Transport Fuel Demand, In T. Sterner (ed.) *International Energy Economics.* London: Chapman & Hall.

Stewart, J.Q. 1948. Demographic Gravitation: Evidence and Application, *Sociometry, 1:* 31-58.

Subba Rao, T. 1981. On the Theory of Bilinear Time Series Models, *Journal of Royal Statistical Society B,* 43:244-255.

Sutton, J.D. (ed.) 1988. *Agricultural Trade and Natural Resources: Discovering the Critical Linkages.* Boulder, CO: Lynne Reiner.

Suwala, N. and W.C. Labys. 1997. *Modeling Transition in the Polish Coal Industry.* WP 190, Natural Resource Economics Program, West Virginia University

Sweeney, J. 1975. *Passenger Car Use of Gasoline: An Analysis of Policy Options*, U.S. Federal Energy Administration, Washington, DC.

Swisko, G.M. 1989. Impacts on Changes in the Nonfuel Mineral Industries to the State and Local Economies of Arizona and Nevada, *Mineral and Materials:* 9-20.

Takayama, T. and G.G. Judge, 1971, *Spatial and Temporal Price and Allocation Models.* Amsterdam: North Holland Publishing Company.

Takayama, T. and W.C. Labys, 1986, Spatial Equilibrium Analysis: Mathematical and Programming Model Formulation of Agricultural, Energy and Mineral Models, In P. Nijkamp (ed.) *Handbook of Regional Economics,* Amsterdam: North-Holland Publishing Company.

Takayama, T. 1979. An Application of Spatial and Temporal Price Equilibrium Model to World Energy Modeling, *Papers of the Regional Science Association,* 41: 43-58.

Takayama, T., and H. Hashimoto. 1984. A Comparative Study of Linear Complementarity Programming Models and Linear Programming Models in Multi-regional Investment Analysis, Paper 1984-1, EPDCS Division, World Bank, Washington, D.C.

Takayama, T., and G.G. Judge. 1964. Equilibrium Among Spatially Separated Markets: Reformulation, *Econometrica,* 32: 510-24.

Takayama, T., and G.G. Judge. 1971. *Spatial and Temporal Price and Allocation Models.* Amsterdam: North-Holland.

Taylor, L. D. 1975. The Demand for Electricity: A Survey, *Bell Journal of Economics,* 6: 74-110.

Taylor, L.D. 1977. The Demand for Energy: A Survey of Price and Income Elasticities• In W.D. Nordhaus (ed.), *International Studies of the Demand for Energy.* Amsterdam: North Holland.

Teisberg, T.J. 1986. A Dynamic Programming Model of the U.S. Strategic Petroleum Reserve, *Bell Journal of Economics,* 17.

Theil, H. 1971. *Principles of Econometrics.* New York: John Wiley.

Theil, H. and K.W. Clements. 1987. *Applied Demand Analysis.* Cambridge, MA: Ballinger.

Thurman, W.N. 1988. Speculative Carryover: An Empirical Examination of the U.S. Refined Copper Market, *RAND Journal of Economics,* 19: 420-437.

Tinbergen, J. 1962. *Shaping the World Economy: Suggestion for an International Economic Policy.* New York: The Twentieth Century Fund.

Tinsley, R.C. 1985. Analysis of Risk Sharing, In C.R. Tinsley (ed.), *Finance for the Minerals Industry,* New York: American Institute of Mining Engineers, 419-425.

Tobey, J. 1990. The Effects of Domestic Environmental Policies on Patterns of World Trade: an Emprical Test, *Kyklos,* 43:191-209.

Tomlin, J. A. 1976. *Programming Guide to LCPL: A Program for Solving Linear Complementarity Problems by Lemke's Method.* Operations Research Center, Stanford University, Stanford.

Tong, H. 1983. *Threshold Models in Non-Linear Time Series Analysis*. Berlin: Springer-Verlag.

Tong, H. and K.S. Lim. 1980. Threshold Autoregression, Limit Cycles and Cyclical Data. *Journal of the Royal Statistical Society*, Series B, 42:245-292.

Torries, T.F. 1998. *Evaluating Mineral Project: Applications and Misperceptions*. Littleton, Co: Society for Mining. Metallurgy and Exploration.

Torries, T.F. and I. Martens. 1985. Nickdata 1985: A Comprehensive Nickel Industry Cost Data Base and Costing Program Nickdata Inc., Toronto.

Torries, T.F. 1983. Competitive Cost Analysis in the Mineral Industries, *Resources Policy*, 12:193-204.

Tramel, T.E., and A.D. Seale, Jr. 1963. Reactive programming: Recent developments, In R.A. King (ed.), *Interregional Competition Research Methods*. Raleigh: University of North Carolina Press.

Trieu, L.H., Savage, E., and G. Dwyer. 1994. A Model of the World Uranium Market, *Energy Policy*, 22: 317-330.

Trozzo, C.L. 1966, Technical Efficiency of the Location of Integrated Blast Furnace Capacity, Ph.D. Thesis, Harvard University.

Tsao, C.S. and R.H. Day, 1971, A Process Analysis Model of the U.S. Steel Industry, *Management Science*, 17: 588-608.

Turvey, R. 1968. *Optimal Pricing and Investment in Electricity Supply*. Cambridge, MA: MIT Press.

Turner, W.C. (ed). 1982. *Energy Management Handbook*. New York: John Wiley& Sons.

Uhler, R.S. and P. Eglington. 1983. The Potential Supply of Crude Oil and Natural Gas Reserves in the Alberta Basin. Monograph prepared for the Economic Council of Canada, Ottawa.

Ulph, A., and M. Folle. 1978. Gains and Losses to Producers for Cartelization of an Exhaustible Resource, C.R.E.S. Working Paper R/WP26. Australian National University, Canberra.

Ulph, A., and M. Folle. 1977-78. Role of Energy Modeling in Policy Formulation, *Energy Systems and Policy*, 2: 311-340.

United Nations. 1994. Price-Forecasting Techniques and their Application to Minerals and Metals in the Global Economy United Nations Publications, New York.

U.S. Central Intelligence Agency (CIA) 1977. • The International Energy Situation Outlook to 1985, CIA, McLean, VA.

U.S. Department of Commerce (DOC). 1974. *Input-Output Structure of the U.S. Economy. 1967. Volume 3: Total Requirements for Detailed Industries*, Washington, DC: U.S. Government Printing Office.

U.S. Department of Energy (DOE). 1978. Midterm Oil and Gas Supply Modeling System Methodology Description, Technical Memorandum TM/ES/79-05. Department of Energy, Washington, D.C.: 43.

Uri, N. 1976. Planning in Public Utilities, *Regional Science and Urban Economics,* 6: 105-25.

Uri, N. 1977. The Impacts of Environmental Regulations on the Allocation and Pricing of Electricity and Energy, *Journal of Environmental Management,* 5: 215-27.

Uri, N., 1976. *Toward an Efficient Allocation of Electric Energy,* Lexington, MA: Heath Lexington Books.

Uri, N., 1979. A Mixed Time Series/Econometric Approach to Forecasting Peak System Load, *Journal of Econometrics,* 9: 155-71.

Uri, N., 1982. *The Demand for Energy and Conservation in the United States.* Contemporary Studies in Energy Analysis and Policy, Vol. I, London: JAI Press Inc.

Verleger, P. K., 1982. *Oil Markets in Turmoil.* Cambridge: Ballinger.

Verleger, P.K., 1993. *Adjusting to Volatile Energy Prices.* Institute for International Economics, Washington, D.C.

Vogely, W.A.(ed.). 1985. *Economics of the Mineral Industries,* New York: American Institute of Mining, Engineers.

Vogely, W.A. (ed.), 1975, *Mineral Materials Modeling.* Washington, DC: Resources for the Future and Johns Hopkins University Press.

Von Neumann, J., and A. Morgenstern. 1944. *Theory of Games and Economic Behavior.* Princeton: Princeton University Press.

Vouyoukas, E. 1996. World Energy Outlook (IEA), In J.B. Lesourd, J. Percebois, and F. Valette (eds.), *Models for Energy Policy.* New York: Routledge.

VP-Expert. 1993. *Rule-Based Expert System Development Tool,* Orinda, CA: WordTech Sytems, Inc.

Waelbroeck, J. L. (ed.). 1976. *The Models of Project LINK.* Amsterdam: North-Holland.

Wagner, H. M. 1969. *Principles of Operations Research.* Englewood Cliffs, NJ: Prentice Hall.

Webb, G., and D.C. Pearce. 1975. The Economics of Energy Analysis, *Energy Policy,* 3: 318-331.

Weston, R.F. and M. Ruth. 1997. A Dynamic Hierarchical Approach to Understanding and Managing Natural Economic Systems, *Ecological Economics*.

Williams, J.C. and B.D. Wright. 1991. *Storage and Commodity Markets*. Cambridge: Cambridge University Press.

Williams, R.H., Larson, E.D., and M.H. Ross. 1987. Materials, Effluence, and Industrial Energy Use, *Annual Review of Energy and the Environment*, 12: 99-144.

Wilson, A.G. 1970. *Entropy in Urban and Regional Modeling*. London: Pion.

Wilson, D., and J. Swisher. 1993. Exploring the Gap: Top-down Versus Bottom-up Analyses of the Cost of Mitigating Global Warming, *Energy Policy*, 21: 249-263.

Winters, L.A. and D. Sapsford. 1990. *Primary Commodity Prices: Economic Models and Policy*. Cambridge: Cambridge University Press.

Wirl, F. 1995. The Exploitation of Fossil Fuels Under the Threat of Global Warming and Carbon Taxes: A Dynamic Game Approach, *Environmental and Resource Economics*, 5:333-352.

Wood, D.O., and M.J. Mason. 1982. Analysis of the Energy Information Administration Coal Supply Model, In J. Quirk, K. Terasawa and D. Whipple (eds.), *Coal Models and Their Use in Government*. New York, Praeger, 37-57.

Wood, D.O., and C. Spierer. 1984. Modeling Swiss Industry Interfuel Substitution in the Presence of Natural Gas Supply Constraints, Energy Lab Report WP84-011, MIT Energy Lab, Cambridge, MA.

Woods, T.J. and H. Vidas. 1983. Projecting Hydrocarbon Production and Cost: an Integrated Approach, *Journal of Energy and Development*, Spring: 267-82.

Workshop on Alternative Energy Strategies. 1977. *Energy: Global Prospects 1985-2000*. New York: McGraw-Hill Publishing Co.

Wright, B.D. and J.C. Williams. 1984. The Welfare Effects of the Introduction of Storage, *Quarterly Journal of Economics*, 104:275-298.

Wright, B.D. and J.C. Williams. 1982. The Economic Role of Commodity Storage, *Economic Journal*, 92: 596-614.

Yaksick, R. (ed.). 1981. Collected Papers from the Workshop on Nonfuel Minerals Demand Modeling, Board on Mineral and Energy Resources, Committee on Nonfuel Mineral Demand Relationships, National Academy of Sciences, Washington, D.C.

Yang, C.W., and W.C. Labys. 1981. Stability of Appalachian Coal Shipments Under Policy Variation, *Energy Journal*, 2: 111-128.

Yang, C.W., and W.C. Labys. 1985. A Sensitivity Analysis of the Linear Complementarity Programming Model: the Case of Appalachian Steam Coal and Natural Gas Markets, *Energy Economics*, 7: 145-52.

Yoshiki-Gravelsins, K.S., Toguri, J.M., and R.T.C. Choo. 1993. Metals Production, Energy, and the Environment, Part II: Environmental Impact, *The Journal of the Minerals and Materials Society*, 45: 23-29.

Yu, E.S.H. and J.C. Jin. 1992. Cointegration Tests of Energy Consumption, Income, and Employment, *Resources and Energy*, 14: 259-66.

Zimmerman, M.B. 1977. Modeling Depletion in a Mineral Industry: the Case of Coal, *Bell Journal of Economics*, 8: 41-65.

Zimmerman, M.B. 1981. *The U.S. Coal Industry: The Economics of Policy Choice.* Cambridge, MA: MIT Press.

INDEX